MATERIALS SCIENCE AND PROCESS TECHNOLOGY SERIES

Editors

Rointan F. Bunshah, University of California, Los Angeles *(Materials Science and Technology)*

Gary E. McGuire, Microelectronics Center of North Carolina *(Electronic Materials and Processing)*

DEPOSITION TECHNOLOGIES FOR FILMS AND COATINGS; Developments and Applications: by *Rointan F. Bunshah et al*

CHEMICAL VAPOR DEPOSITION FOR MICROELECTRONICS; Principles, Technology, and Applications: by *Arthur Sherman*

SEMICONDUCTOR MATERIALS AND PROCESS TECHNOLOGY HANDBOOK; Very Large Scale Integrated Circuits (VLSIC) and Ultra Large Scale Integrated Circuits (ULSIC): edited by *Gary E. McGuire*

SOL-GEL TECHNOLOGY; Principles, Developments and Applications: edited by *Lisa C. Klein*

HYBRID MICROCIRCUIT TECHNOLOGY HANDBOOK; Materials, Processes, Design, Testing and Production: by *James J. Licari* and *Leonard R. Enlow*

HANDBOOK OF THIN FILM DEPOSITION PROCESSES AND TECHNIQUES; Principles, Methods, Equipment and Applications: edited by *Klaus K. Schuegraf*

Related Titles

ADHESIVES TECHNOLOGY HANDBOOK: by *Arthur H. Landrock*

HANDBOOK OF THERMOSET PLASTICS: edited by *Sidney H. Goodman*

HANDBOOK OF CONTAMINATION CONTROL IN MICROELECTRONICS; Principles, Applications and Technology: edited by *Donald L. Tolliver*

CHEMICAL VAPOR DEPOSITION FOR MICROELECTRONICS

Principles, Technology, and Applications

by

Arthur Sherman

Varian Research Center
Palo Alto, California

np

NOYES PUBLICATIONS
Park Ridge, New Jersey, U.S.A.

Library of Congress Catalog Card Number: 87-11277
ISBN: 0-8155-1136-1
Printed in the United States

Published in the United States of America by
Noyes Publications
Mill Road, Park Ridge, New Jersey 07656

10 9 8 7 6 5 4 3 2

Library of Congress Cataloging-in-Publication Data

Sherman, Arthur.
Chemical vapor deposition for microelectronics.

Bibliography: p.
Includes index.
1. Vapor-plating. 2. Integrated circuits--Design and construction. I. Title.
TS695.S54 1987 621.381'7 87-11277
ISBN 0-8155-1136-1

Preface

The objective of the present text on Chemical Vapor Deposition (CVD) is to present a unified picture of an interdisciplinary field. There are many references that deal in great detail with limited aspects of the subject, but none that encompass all elements. For example, early CVD reactors tended to operate at atmospheric pressures, and many researchers studied the fluid dynamic nature of such systems (recirculating flows, buoyancy effects, etc.). Recently, low pressure systems have become of interest, and by and large, the fluid dynamic character of the reactor flow is not studied in detail. Such an approach is acceptable for initial operation of these systems. However, as demands on them continue to grow, it becomes necessary to again consider the fluid dynamics. Similarly, many cold-wall as well as hot-wall reactor systems have been used commercially. Again, as these systems are pushed to their limits, it becomes apparent that there are fundamental differences in their operation. It is doubtful that such differences will be clarified until researchers include fluid dynamics with gas phase kinetics and with surface kinetics in their studies. To summarize, CVD is the study of the flow of reactive gas mixtures with heterogeneous surface reactions. Because of the inordinate complexity of the problem, most studies of the subject have been empirical. It is the author's hope that the present text will encourage more studies of CVD phenomena from first principles.

In the first chapter, we consider the fundamental nature of the thermally-induced CVD. Initially, we consider the behavior of CVD reactions under the assumption of chemical equilibrium. Much useful information can be derived by this technique, especially for very complex chemical systems where several different solid phases can be deposited. In order to extend our understanding of CVD, it is necessary to consider reacting gas flows where the rates of chemical reactions are finite. Therefore, the next subject considered is the modeling of CVD flows, including chemical kinetics. Depending on processing conditions, the film being deposited may be amorphous, polycrystalline, or epitaxial, so the morphology of deposited films is discussed briefly. Finally, the thermal CVD reactor configurations that have been typically used in research and development are reviewed.

In addition to thermally-created CVD films, much work has been done using glow discharges to modify the deposition. Therefore, Chapter 2 reviews the fundamentals of plasma-enhanced CVD (PECVD). Initially, the basic character of a plasma is covered. Then we discuss the influence of the reactor configuration on the plasma behavior and PECVD deposition. The two major PECVD reactor systems are reviewed, and then several new concepts are considered.

The next three chapters review the deposition of thermally-induced dielectric films (Chapter 3) and metallic conducting films (Chapter 4), as well as plasma-enhanced films of either type (Chapter 5). The many chemical systems employed to create these films are considered, and the nature of the resulting films is presented. Films studied are silicon dioxide, silicon nitride, polysilicon, epitaxial silicon, the refractory metal silicides, tungsten and aluminum.

Chapter 6 is devoted to typical commercially-available CVD reactor systems, including cold-wall and hot-wall systems. Several new commercial reactors are also reviewed.

Finally, Chapter 7 covers methods commonly used for film evaluation. The first portion covers techniques for assessing the physical nature of the films produced, while the latter portion reviews methods of chemical analysis of thin films.

The author wishes to express his gratitude to Varian Associates for providing the necessary facilities for the preparation of the manuscript, and to Mrs. Nancy Anderson for her patient and careful typing of the text. Finally, I must thank Drs. G.J. Reynolds, C.B. Cooper III, J.A. Fair and S.B. Felch for their assistance in reviewing the manuscript.

Palo Alto, California
July, 1987

Arthur Sherman

NOTICE

To the best of the Publisher's knowledge the information contained in this publication is accurate; however, the Publisher assumes no liability for errors or any consequences arising from the use of the information contained herein. Final determination of the suitability of any information, procedure, or product for use contemplated by any user, and the manner of that use, is the sole responsibility of the user.

The book is intended for information only. The reader is warned that caution must always be exercised when dealing with hazardous materials, and expert advice should be obtained at all times when implementation is being considered.

Contents

1

Fundamentals of Thermal CVD

1.1 INTRODUCTION

Chemical vapor deposition (CVD) is a process where one or more gaseous species react on a solid surface and one of the reaction products is a solid phase material. For example, consider the pyrolysis of silane (SiH_4) on a hot surface. When a silane molecule strikes a surface, it can either be reflected or adsorbed. If it is adsorbed, and the temperature is high enough to promote its decomposition, it may decompose into Si and H_2 with the latter going back into the gas phase. The silicon left behind can build up as a thin solid film. Similar reactions occur where two compounds adsorb onto a surface and react there, leaving behind a solid phase. If all products of the surface reaction are gaseous, the process is called heterogeneous gas phase catalysis. Much research has been done in this field which will eventually be useful in understanding CVD.

The several steps that must occur in every CVD reaction are as follows:

(1) Transport of reacting gaseous species to the surface.

(2) Adsorption, or chemisorption, of the species on the surface.

(3) Heterogeneous surface reaction catalyzed by the surface.

(4) Desorption of gaseous reaction products.

(5) Transport of reaction products away from the surface.

In all CVD processes, we are dealing with the change from one state (i.e., the initial, low-temperature reactant gases) to a later one (i.e., the final state with some solid phase and product gases) in time. Since any practical commercial process must be completed quickly, the rate with which one proceeds from the initial to the final state is important. This rate will depend on chemical kinetics (reaction rates) and fluid dynamic transport phenomena. Therefore, in order to clearly understand CVD processes, we will not only examine chemical thermodynamics (Section 1.2), but also kinetics and transport (Section 1.3).

Two types of CVD systems can be considered. One is a closed system into which a finite quantity of reactant gas is introduced, such as shown in Figures 1A and 1B. Initially, the silane (SiH_4) is introduced at a low temperature (T_o). Silane will then diffuse to the hot wall through a concentration gradient layer, adsorb on the walls, dissociate there and leave solid silicon behind while H_2 diffuses back into the gas. After a finite time, an equilibrium is reached where no more silicon is deposited.

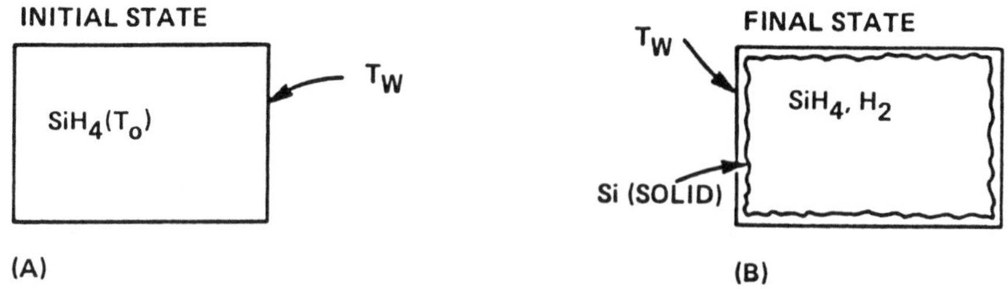

Figure 1: Closed CVD system.

The second type is the open system. Actually, this is the most common system used, and the one we will be primarily studying in this text. We deal here with a flowing system such as shown schematically in Figure 2.

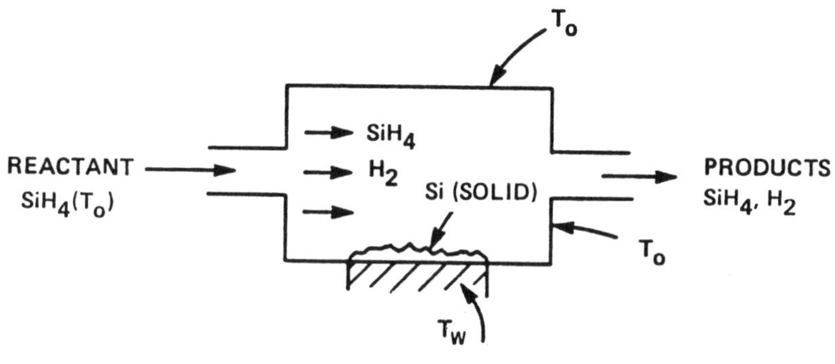

Figure 2: Open flow CVD system.

In this situation, a film is grown on the hot surface (T_W), and its thickness will increase without limit as long as fresh reactants are provided and products can be removed. The gas state will be in quasiequilibrium far from the hot surface and in a strongly nonequilibrium condition close to it. The change from one to the other will occur across a boundary layer where temperature, velocity, and species concentration vary rapidly. The behavior of this boundary layer will be determined by gas transport properties such as viscosity, thermal conductivity, as well as gas-phase kinetics and diffusion coefficients. So, even if the kinetics at the surface are very fast, we must deal with quasiequilibrium phenomena where gas conditions vary rapidly over short distances.

In the analysis of CVD reactions, it is important to recognize the rates of the various processes. The slowest rate will be controlling, and which one is the slowest or fastest can depend on gas as well as surface conditions. For example, surface reactions may be fast at high surface temperatures. In this case, the CVD process will tend to be limited by the rate at which reactants can get to the surface or products leave it. For this situation, the fluid dynamic boundary layer phenomena will govern the deposition rate. On the other hand, at low pressures diffusion is very rapid and the rate at which surface reactions proceed will tend to govern the deposition rate. Alternatively, low surface temperatures will have low reaction rates, and this will govern no matter how much material diffuses to the surface.

The essential issues that one is concerned with in all CVD processes are:

(1) Nature of solid deposit given particular reactants.

(2) Rate of deposition of solid film.

(3) Uniformity of deposition over extended surfaces.

(4) Morphology of solid film.

In the remainder of this chapter, some of the basic ideas governing these issues will be covered.

1.2 CHEMICAL EQUILIBRIUM

Although CVD processes inherently involve rapid changes, it is useful to examine the limiting case of long reaction times for insights into the nature of the films that can be deposited. To do this, we examine the final equilibrium state for the reactions of interest, which will depend on the initial reactant gas composition and the final pressure and temperature.

The problem we are addressing here is: what is the gas phase composition of a mixture of gases under specified conditions of pressure and temperature, where as much time as is necessary is allowed for the gases to equilibrate? If there is a change of phase as one proceeds from one equilibrium state to another (i.e., solid silicon film forming on the container walls), then this has to be accounted for as well.

1.2.1 Law of Mass Action

Historically, the state of reaction at chemical equilibrium was evaluated for fairly simple reactions, with only a few species, from the "Law of Mass Action."[1] In recent years, high-temperature reactions, including many possible species (as many as 20 or more), have become of interest and newer techniques suitable for numerical solution on high-speed digital computers have been developed.[2] Initially, we will discuss chemical equilibrium from the vantage point of the "Law of Mass Action." It states that the rate at which a chemical reaction proceeds is proportional to the "active" masses of the reacting substances. The active mass for a mixture of ideal gases is the number density of each reacting species, or for a given temperature, it can be represented by its partial pressure.

Consider a typical reaction,

$$\nu_a A + \nu_b B \rightleftarrows \nu_c C + \nu_d D$$

Then the "Law of Mass Action" states that

$$\frac{(p_c)^{\nu_c} (p_d)^{\nu_d}}{(p_a)^{\nu_a} (p_b)^{\nu_b}} = K_p (T) \qquad (1)$$

where p is the partial pressure, ν is the stoichiometric coefficient, and K_p is the equilibrium constant which is a function of the temperature alone.

As a simple example of how this relation may be used to establish the equilibrium composition of reacting gases, consider the dissociation of nitrogen tetroxide when its temperature is increased from room temperature to some elevated value.

$$N_2O_4 \rightleftarrows 2NO_2$$

Equation (1) gives

$$K_p(T) = \frac{p^2_{NO_2}}{p_{N_2O_4}} \qquad (2)$$

but we also know that

$$p_{NO_2} + p_{N_2O_4} = p \qquad (3)$$

where p is the total pressure. Equations (2) and (3) can be solved for p_{NO_2} and $p_{N_2O_4}$ in terms of $K_p(T)$ and p.

For reactions in which the number of molecules do not change during the reaction, the amount of each reactant decomposed at equilibrium will be independent of the total pressure. Consider the reaction

$$N_2 + O_2 \rightleftarrows 2NO$$

In this case

$$K_p(T) = \frac{p^2_{NO}}{p_{N_2} p_{O_2}} \qquad (4)$$

and we know

$$p = p_{NO} + p_{N_2} + p_{O_2} \tag{5}$$

if

$$p_{N_2} = p_{O_2}$$

then Equations (4) and (5) become

$$K_p(T) = \frac{p_{NO}^2}{p_{N_2}^2} \tag{6}$$

and

$$p = p_{NO} + 2p_{N_2} \tag{7}$$

Solving Equations (6) and (7) for p_{NO} gives

$$p_{NO} = \frac{\sqrt{K_p}}{2 + \sqrt{K_p}} \cdot p \tag{8}$$

Now, assume we start with A moles of N_2 plus O_2. We then heat this mixture to temperature T, at which point x molecules of each N_2 and O_2 have reacted to form NO. We can then write the partial pressure of NO formed as

$$p_{NO} = \frac{2x}{A} p \tag{9}$$

Comparing Equations (8) and (9), and recalling that $K_p = K_p(T)$, we recognize that the extent of the reaction (x) is independent of the total pressure p, and depends only on the temperature through K_p.

The behavior of a reaction when an inert gas or an excess of reactants are added is of interest. We can understand this situation if we rewrite the partial pressures in Equation (1) as

$$p_a = \frac{n_a}{n} p \tag{10}$$

where n_a/n is the mole fraction of species a. Then Equation (1) becomes

$$K_p = \frac{(n_c)^{\nu_c} (n_d)^{\nu_d}}{(n_a)^{\nu_a} (n_b)^{\nu_b}} \left(\frac{p}{n}\right)^{\Delta\nu} \tag{11}$$

when $\Delta\nu$ is positive (more molecules on right-hand side than left-hand side of reaction equation), the addition of an inert gas, while keeping p the same, results in an increase in n which must be compensated by a corresponding decrease in n_a and n_b. In other words, the reaction will shift to the right. For a negative $\Delta\nu$, the reverse is true. Also, an excess of one of the reactants will have the same effect. For example, consider the dissociation of PCl_5.

$$PCl_5 \rightleftarrows PCl_3 + Cl_2 \quad ,$$

where $\Delta\nu = 1$. If we add Cl_2 or any inert gas to the PCl_5 and keep p constant, then n will increase and the reaction will shift toward the products. Said another way, if one has dilute phosphorus pentachloride (in an inert gas or chlorine mixture), then more of it will dissociate than if we had 100% PCl_5. We will promote the dissociation reaction.

When a heterogeneous reaction is considered, the partial pressures included in the "Law of Mass Action," Equation (1), are only those for the gaseous reactants. For example, when silicon is deposited on a surface due to silane pyrolysis, we have

$$SiH_4(g) \rightleftarrows Si(s) + 2H_2(g)$$

and

$$\text{(12)} \qquad \left(\frac{p_{SiH_4}}{p^2_{H_2}}\right)^{-1} = K_p\ (T)$$

The partial pressure of Si does not appear because it is in equilibrium with the solid, and at a given temperature is a constant regardless of pressure. Said another way, the activity of silicon is 1 when it is present as a solid in pure form.

The equilibrium constant for each reaction, as a function of temperature, was originally determined experimentally. However, the standard free energy change is related to the equilibrium constant by[1]

$$\text{(13)} \qquad \Delta F^\circ = -RT\ \ell n\ K_p \quad ,$$

where R is the gas constant. The convention is that the ΔF° value for elements is zero, and for other compounds these values are tabulated as functions of temperature. Adding the ΔF° values for each compound in a reaction gives the value for the reaction as a whole. For example, consider the water gas reaction

$$CO_2 + H_2 \rightleftarrows H_2O + CO \quad .$$

Now,

$$\Delta F^\circ_{CO_2} = -94.45\ \text{KCal/mole}$$

$$\Delta F^\circ_{H_2} = 0$$

$$\Delta F^\circ_{H_2O} = -54.65 \text{ KCal/mole}$$

$$\Delta F^\circ_{CO} = -33.00 \text{ KCal/mole} .$$

Then, ΔF° for the overall reaction is -33.00 + (-54.65) – (-94.45), or 6.8 Kcal per mole. Therefore, the equilibrium constant for this reaction can be calculated from Equation (13) as

$$K_p = e^{-\frac{6,800}{RT}} .$$

1.2.2 Reactions with Multiple Species

When there are many species involved, the problem becomes much more complex and species partial pressures have to be calculated by approximate numerical techniques. As an illustration, consider the Si-Cl-H system[3] with only eight gaseous species (H_2, HCl, SiH_4, SiH_3Cl, SiH_2Cl_2, $SiHCl_3$, $SiCl_4$ and $SiCl_2$) allowed, where the deposition of solid Si on the surface of the container must be allowed for.

It must also be made clear that the accuracy of this approach depends on the skill with which the significant species are selected. For example, Si_2H_6 may be a significant species in this reaction, but it has been neglected.

To treat this problem along the lines we have been discussing, we choose six reactions as follows:

(14)

$$Si(s) + 4HCl(g) \rightleftarrows SiCl_4(g) + 2H_2(g)$$

$$K_p(1) = \frac{p_{SiCl_4}\, p^2_{H_2}}{p^4_{HCl}}$$

(15)

$$Si(s) + 3HCl(g) \rightleftarrows SiCl_3H(g) + H_2(g)$$

$$K_p(2) = \frac{p_{SiCl_3H}\, p_{H_2}}{p^3_{HCl}}$$

(16)

$$Si(s) + 2HCl(g) \rightleftarrows SiCl_2H_2(g)$$

$$K_p(3) = \frac{p_{SiCl_2H_2}}{p^2_{HCl}}$$

(17)
$$Si(s) + HCl(g) + H_2 \rightleftarrows SiClH_3(g)$$
$$K_p(4) = \frac{p_{SiClH_3}}{p_{HCl}p_{H_2}}$$

(18)
$$Si(s) + 2HCl(g) \rightleftarrows SiCl_2(g) + H_2(g)$$
$$K_p(5) = \frac{p_{SiCl_2}\,p_{H_2}}{p^2_{HCl}}$$

(19)
$$Si(s) + 2H_2(g) \rightleftarrows SiH_4(g)$$
$$K_p(6) = \frac{p_{SiH_4}}{p^2_{H_2}}$$

A seventh equation is derived from the fact that the sum of the partial pressures must equal the total pressure (in this example, 1 atmosphere).

(20)
$$p_{SiCl_4} + p_{SiCl_3H} + p_{SiCl_2H_2} + p_{SiClH_3} + p_{SiCl_2} + p_{SiH_4} + p_{HCl} + p_{H_2} = 1$$

A final (eighth) relationship is obtained when we specify the Cl/H ratio. For this system, Si may leave the gas phase in going from one equilibrium to another (deposit as a thin film), but the Cl and H will remain the same. Therefore, we can write

(21)
$$\frac{Cl}{H} = \frac{4p_{SiCl_4} + 3p_{SiCl_3H} + 2p_{SiCl_2H_2} + 2p_{SiCl_2} + p_{SiClH_3} + p_{HCl}}{2p_{H_2} + p_{SiCl_3H} + 2p_{SiCl_2H_2} + 3p_{SiClH_3} + p_{HCl} + 4p_{SiH_4}}$$

These eight equations [Equations (14) through (21)] can now be solved for the eight species partial pressures. Since they are highly nonlinear, they were solved by a numerical iterative procedure on a high-speed digital computer.[3] The Free Energies of Formation were evaluated from available thermodynamic data.[4] Additional sources of such data are generally available.[5]

The results of this calculation are presented in Table 1. For each value of Cl/H and temperature, there will be a calculated value Si/Cl. If the original value of Si/Cl (at room temperature) is larger than this, then Si deposition will occur in going from the initial state to the final one. If the original value is less than the final state, etching of Si will occur. Clearly, since we are comparing equilibrium states, there is no way we can evaluate deposition/etch rates. For this, we will have to look at the kinetics of this situation.

The Si/Cl ratio can be calculated from

$$\frac{Si}{Cl} = \frac{p_{SiCl_4} + p_{SiCl_3H} + p_{SiCl_2H_2} + p_{SiCl_2} + p_{SiClH_3} + p_{SiH_4}}{4p_{SiCl_4} + 3p_{SiCl_3H} + 2p_{SiCl_2H_2} + 2p_{SiCl_2} + p_{SiClH_3} + p_{HCl}} \tag{22}$$

Table 1: Equilibrium Partial Pressures of Species (H-Cl-Si System, Atmospheres)[3]

T, °K	Cl/H	H_2	HCl	$SiCl_4$	$SiHCl_3$
1000	10^{0}	6.28×10^{-1}	1.52×10^{-2}	2.74×10^{-1}	8.01×10^{-2}
1000	10^{-1}	9.39×10^{-1}	1.07×10^{-2}	3.03×10^{-2}	1.88×10^{-2}
1000	10^{-2}	9.90×10^{-1}	5.40×10^{-3}	1.74×10^{-3}	2.26×10^{-3}
1000	10^{-3}	9.98×10^{-1}	1.67×10^{-3}	1.55×10^{-5}	6.58×10^{-5}
1200	10^{0}	6.0×10^{-1}	5.80×10^{-2}	2.61×10^{-1}	7.36×10^{-2}
1200	10^{-1}	9.18×10^{-1}	3.97×10^{-2}	2.44×10^{-2}	1.54×10^{-2}
1200	10^{-2}	9.83×10^{-1}	1.50×10^{-2}	4.30×10^{-4}	7.72×10^{-4}
1200	10^{-3}	9.98×10^{-1}	1.98×10^{-3}	1.26×10^{-7}	1.74×10^{-6}
1400	10^{0}	5.37×10^{-1}	1.38×10^{-1}	2.16×10^{-1}	5.99×10^{-2}
1400	10^{-1}	8.81×10^{-1}	8.63×10^{-2}	1.22×10^{-2}	8.89×10^{-3}
1400	10^{-2}	9.81×10^{-1}	1.85×10^{-2}	2.07×10^{-5}	7.84×10^{-5}
1400	10^{-3}	9.98×10^{-1}	1.98×10^{-3}	2.66×10^{-9}	9.56×10^{-8}

T, °K	SiH_2Cl_2	SiH_3Cl	SiH_4	$SiCl_2$
1000	2.70×10^{-3}	5.05×10^{-5}	2.55×10^{-7}	1.99×10^{-4}
1000	1.34×10^{-3}	5.33×10^{-5}	5.70×10^{-7}	6.62×10^{-5}
1000	3.39×10^{-4}	2.83×10^{-5}	6.34×10^{-7}	1.59×10^{-5}
1000	3.22×10^{-5}	8.78×10^{-6}	6.45×10^{-7}	1.50×10^{-6}
1200	2.92×10^{-3}	5.95×10^{-5}	3.43×10^{-7}	5.01×10^{-3}
1200	1.37×10^{-3}	6.23×10^{-5}	8.02×10^{-7}	1.53×10^{-3}
1200	1.94×10^{-4}	2.52×10^{-5}	9.22×10^{-7}	2.03×10^{-4}
1200	3.38×10^{-6}	3.37×10^{-6}	9.49×10^{-7}	3.49×10^{-6}
1400	2.63×10^{-3}	5.75×10^{-5}	3.64×10^{-7}	4.62×10^{-2}
1400	1.03×10^{-3}	5.88×10^{-5}	9.77×10^{-7}	1.10×10^{-2}
1400	4.71×10^{-5}	1.40×10^{-5}	1.21×10^{-6}	4.52×10^{-4}
1400	5.44×10^{-7}	1.53×10^{-6}	1.25×10^{-6}	5.13×10^{-6}

Referring to Table 1, we find at T = 1000°K and Cl/H = 1 a value of Si/Cl = 0.263. If we had started with dichlorosilane, SiH_2Cl_2, we would have had Si/Cl = 0.500 initially, and deposition would have occurred. If instead we had started with silicon tetrachloride, $SiCl_4 + 2H_2$, the initial value would have been Si/Cl = 0.250 and etching would have been the result. Then, if we add much more H_2 to the original mixture (i.e., $SiCl_4 + 2000\,H_2$), we would have Cl/H = 10^{-3} as well as Si/Cl = 0.250, and would find Si/Cl = 0.0618. So, using $SiCl_4$ at 1000°K, we can go from etching to deposition, depending on how much H_2 we add to the mixture. Figure 3 shows the temperature variation of the deposition/etch regimes as a function of the partial pressure of silicon tetrachloride which is mixed with hydrogen. Clearly, increasing the gas mixture temperature can lead to etching rather than the desired deposition unless the $SiCl_4$ is heavily diluted with H_2.

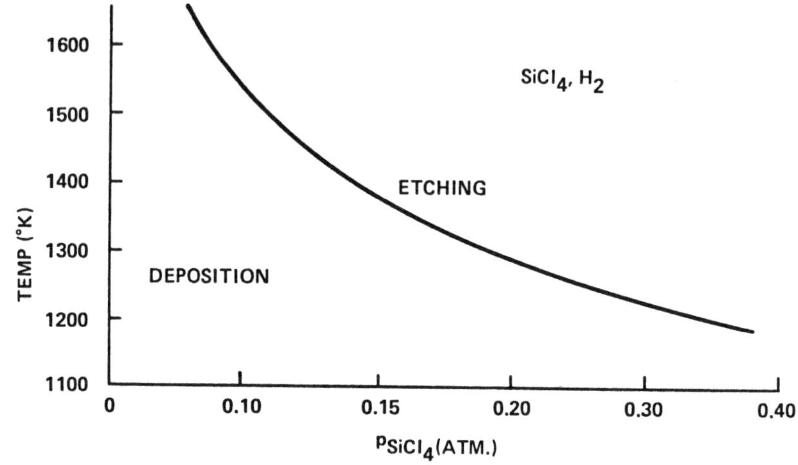

Figure 3: Boundary between etching and deposition in a $SiCl_4$ and H_2 mixture at one atmosphere.

1.2.3 Minimization of Gibbs Free Energy

As noted earlier, when chemical reactions become more complex (i.e., at high temperatures, 20 or more species are important), it becomes increasingly more difficult to calculate the equilibrium species concentrations. Initially, we have discussed the equilibrium constant approach to such calculations. This required a priori knowledge of the significant species developing in the reaction (phases as well), and we were required to write out the appropriate reaction equations. In reality, it should not be necessary to specify these reactions, since changing from one thermodynamic state to another should be independent of reaction paths involved. Also, it is not always possible to correctly choose which species will be significant or what phases will be formed. To deal with these more complex situations, another approach has been developed to calculate number densities at thermodynamic equilibrium for arbitrarily large numbers of species.

As is well known from thermodynamics,[6] a system will be in equilibrium when the Gibbs Free Energy is at a minimum. Therefore, all that is necessary is to express the Gibbs Free Energy in terms of the degree of completion of the reaction, and then minimize that function. The Gibbs Free Energy can be expressed as follows,[7]

$$G = \sum_{i=1}^{m} \left(n_i(g)\ \Delta F^o_{f_i}(g) + RT\ell nP + RT\ell n \frac{n_i(g)}{N(g)} \right) + \sum_{i=1}^{s} n_i(c)\ \Delta F^o_{f_i}(c) \tag{23}$$

where

m = number of gaseous species
s = number of solid phase species
$n_i(g)$ = number of moles of gaseous species i
$n_i(c)$ = number of moles of condensed species i
$N(g)$ = total number of moles of gaseous species
$\Delta F^\circ_{f_i}(g)$ = free energy of formation of gaseous species i
$\Delta F^\circ_{f_i}(c)$ = free energy of formation of solid species i
P = total pressure
T = temperature
R = gas constant

and values of the n_i's have to be found that minimize G, subject to the mass balance constraint. That is, the number of atoms of a particular element must be conserved. First, let us define a_{ij} as the formula numbers specifying the number of atoms of element j in a molecule of species i. For example, for the species CH_4, we define i = 1 and let j = 1 represent C and j = 2 represent H, so that $a_{11} = 1$ and $a_{12} = 4$ for this case. Then the constraint can be written as q equations:

$$\sum_{i=1}^{m} a_{ij}(g)\ n_i(g) + \sum_{i=1}^{s} a_{ih}(c)\ n_i(c) = b_j \qquad (j = 1, 2, \ldots, q) \tag{24}$$

where b_j is the number of moles of element j in the original mixture, and there are q elements in the system.

Then the solution to the problem of determining the equilibrium state at thermodynamic equilibrium reduces to one of finding the minimum in the function G subject to the constraints of Equation (24). There are a number of numerical techniques for the solution of this minimization problem,[7,8] but rather than review the details, we will simply describe some of the results of such calculations.

As noted earlier, the great value of the free energy minimization technique is that one can consider completely new systems and not exclude any species. In this way, one identifies all potentially important species. As an illustration, we can consider the very complex Nb-Ge-Cl-H system for which equilibrium calculations have been made.[9] The motivation for the study was to determine conditions suitable to the deposition of Nb_3Ge, which has a high superconducting transition temperature. The initial gaseous reactants were $NbCl_5$, $GeCl_4$ and H_2. Gaseous and condensed species that were considered are shown in Table 2. Clearly, a system with 17 gaseous species and 11 condensed species is too complex to approach by any other technique.

Table 2: Chemical Species and Data Used in the Thermodynamic Calculations for Nb-Ge-Cl-H System, After Wan[9]

Gaseous Species	ΔH°_{f1200K} (kcal/mole)	S°_{1200K} (cal/mole-deg)
Nb	117.65	53.80
Ge	89.88	49.64
H_2	0	41.03
Cl_2	0	65.38
HCl	-22.65	54.65
Cl	29.80	46.93
H	53.38	34.31
GeCl	31.43	71.97
$GeCl_2$	-41.25	90.67
$GeCl_4$	-117.95	117.53
GeH_4	18.74	74.69
GeH_3Cl	-15.00	88.29
GeH_2Cl_2	-52.00	102.5
$GeHCl_3$	-86.05	109.92
Ge_2Cl_6	-135.0	161.8
$NbCl_4$	-131.2	126.48
$NbCl_5$	-164.0	150.89
Condensed Species*		
$NbCl_{2.33}$	-108.76	61.7
$NbCl_{2.67}$	-123.73	64.54
$NbCl_3$	-133.55	69.85
$NbCl_{3.13}$	-137.68	72.33
$NbGe_{0.33}$	- 12.0	22.9
$NbGe_{0.54}$	- 14.6	26.6
$NbGe_{0.67}$	- 15.4	28.5
$NbGe_2$	- 20.8	49.9
Nb	0	17.51
Ge	0	15.91
$NbH_{0.67}$	- 10.0	18.71

*Each formula is written to contain one niobium, except for pure Ge.

The phase diagram for condensed phases in this system is shown in Figure 4. Apparently, there is a fairly narrow regime where Nb_3Ge can be obtained without contamination from other solid species. In Figure 5, the gas composition is shown as a function of temperature, pressure, and source gas composition.

Other systems can be treated in a similar manner such as the Ti-B-Cl-H system,[10] and of course the Si-H-Cl system has been exhaustively studied because of the commercial importance of silicon epitaxial films used for integrated circuit fabrication.[11]

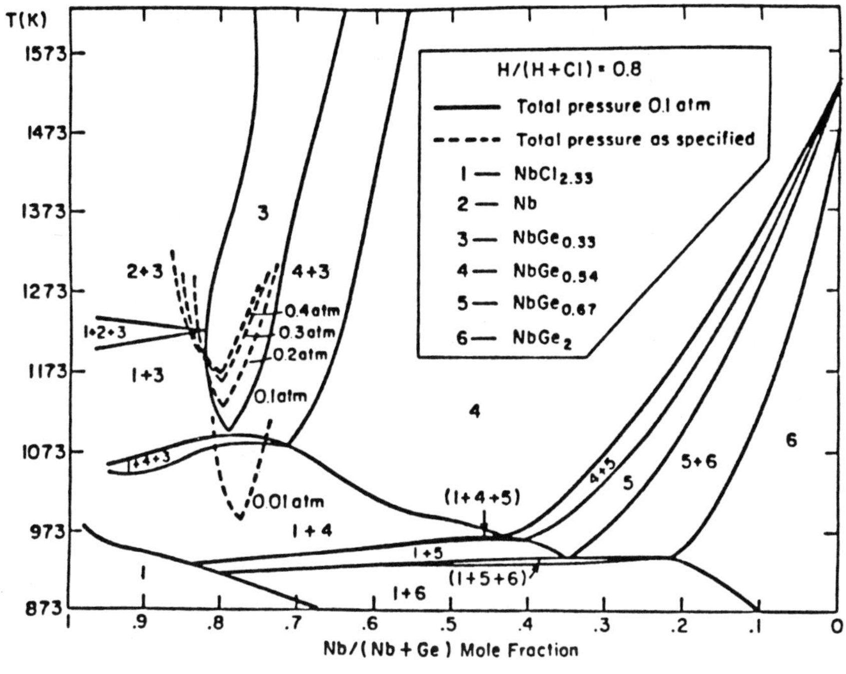

Figure 4: Equilibrium CVD phase diagram for the Nb-Ge-H-Cl system. The diagram was constructed from thermodynamic calculation results and depicts the condensed phases which form as a function of experimental variables. The Nb/(Nb+Ge) values are reactant gas concentrations. After Wan.[9]

1.3 MODELING OF FLOW AND CHEMICAL KINETICS

As noted earlier, our principal interest will be in an open, flowing CVD system. In order to correctly interpret the phenomena occurring in such systems, it will be necessary to study chemically reacting gas flows with nonuniform flow and temperature fields. And, of course, we will have to understand the surface reactions that lead to the solid film formation. Within the reacting gas, we will consider homogeneous gas phase reactions. At the surface, we have

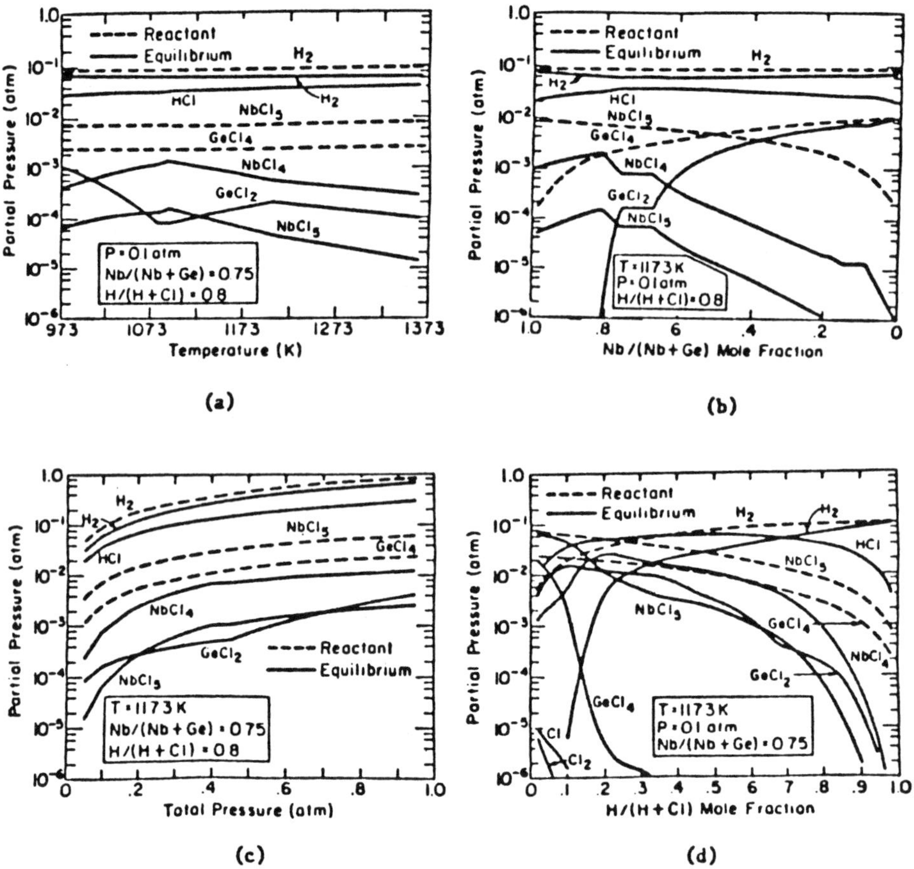

Figure 5: Equilibrium gas compositions as functions of (a) temperature, (b) total pressure, (c) Nb/(Nb+Ge) mole fraction, and (d) H/(H+Cl) mole fraction in the reactant gas. After Wan.[9]

to work with heterogeneous surface reactions. In special situations, we may make use of chemical equilibrium arguments to evaluate deposition phenomena, but in general, an accurate representation of the process will require consideration of chemical kinetics.

1.3.1 Diffusion vs. Surface Controlled Deposition

Before considering the full problem in all its complexity, it will be useful to understand the concept of "surface" or "diffusion" controlled CVD. At one extreme, with very low pressures (large diffusion coefficients) and low surface temperatures, there is a large flux of reactants to the surface where they react slowly so that there is an oversupply of reactants waiting to be consumed. This would be considered the "surface" controlled regime where the rate of deposition is more controlled by the surface temperature than by the details of what is occurring in the bulk gas.

At the other extreme, we have higher pressures (smaller diffusion coefficients) and high surface temperatures. Now, any molecule that can make it to the surface will react rapidly, so that the deposition rate will be more limited by diffusion through the gas adjacent to the surface. Since diffusion is weakly dependent on temperature, this type of "diffusion" controlled regime tends to be relatively insensitive to surface temperature.

It will be instructive to quantify this phenomena by approximate arguments, although actual predictions will depend on more comprehensive models.[12,13] Initially, we recognize that an understanding of fluid dynamic behavior is essential to any attempt to describe these phenomena. In particular, we have the concept of a boundary layer. When a fluid (gas or liquid) flows adjacent to a solid surface, the velocity of the fluid at the surface must be zero. The region in which the fluid velocity changes from its normal value to the zero value on the surface is referred to as a boundary layer (see Figure 6). For high velocity flows, the thickness of this transition region, δ, can be quite small and is approximately proportional to the inverse of the square root of the Reynold's number. Or,

$$\delta = \sqrt{\frac{x}{R_e}} \tag{25}$$

where: $R_e = \frac{\rho u x}{\mu}$

ρ = mass density

u = flow velocity

x = distance in flow direction

μ = viscosity

Similar boundary layers (transition regions) exist for temperature as well as species concentrations.

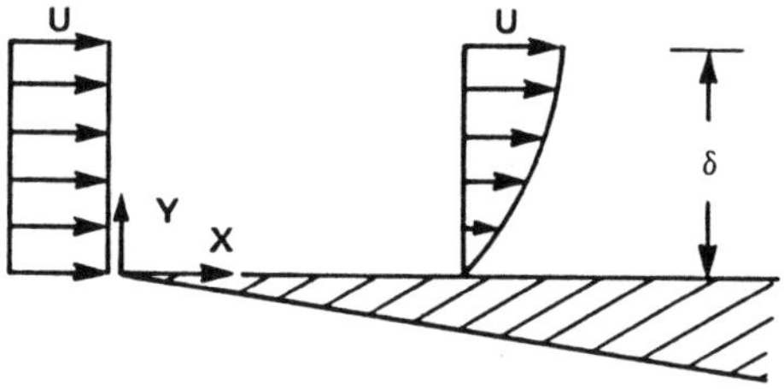

Figure 6: Fluid dynamic boundary layer.

Assume we have a mixture of ideal reacting gases, and a concentration boundary layer. Then the diffusion of species i from the body of the gas to the solid surface is governed by Fick's Law[1]

$$J = D \frac{d\rho}{dy} \tag{26}$$

where D is the diffusion coefficient for diffusion of species i in the mixture, ρ is the local density of species i, and J is the mass flux. Alternately, we can express J in terms of a pressure gradient from the perfect gas law, $\rho = p/RT$.

$$J = \frac{D}{RT} \frac{dp}{dy} \tag{27}$$

or assuming p varies linearly across the boundary layer,

$$J = \frac{D}{RT} \frac{\Delta p}{\Delta y} = \frac{D}{RT} \frac{(p_b - p_s)}{\delta} \tag{28}$$

where: p_b = partial pressure of i in gas
p_s = partial pressure of i at surface

When deposition is controlled by diffusion, Equation (28) shows that variations in boundary layer thickness, δ, influence the mass flux due to diffusion and thereby, the deposition rate. In practical CVD reactors, boundary layer thicknesses can vary so that thickness uniformity of deposition can be poor unless this phenomena is recognized and corrected.

For example, when diffusion-controlled deposition is done in a tube such as that shown in Figure 7, the boundary layer grows and there is an exponential decrease in deposition thickness as x increases.[14] To correct for this phenomena, the susceptor is tilted up, as shown in Figure 8. As the susceptor is tilted up, the velocity above it increases due to the constriction in the channel. The increased velocity increases the Reynold's number, which thins the boundary layer so that the deposition rate goes up. The net effect is to maintain a relatively uniform deposition along x.

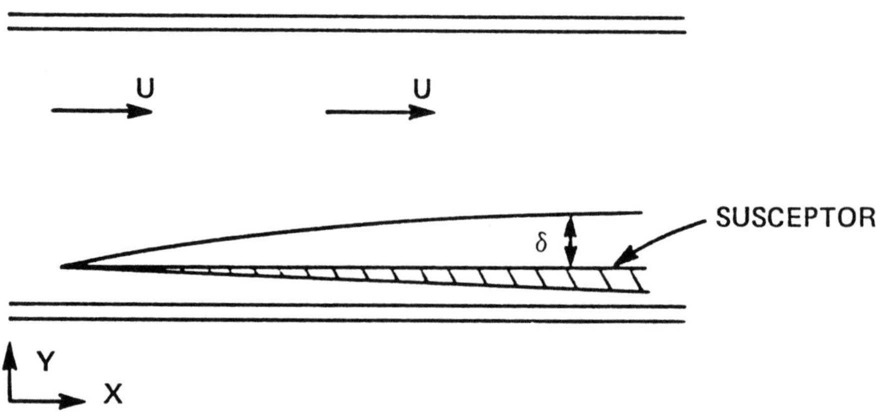

Figure 7: Axial flow reactor.

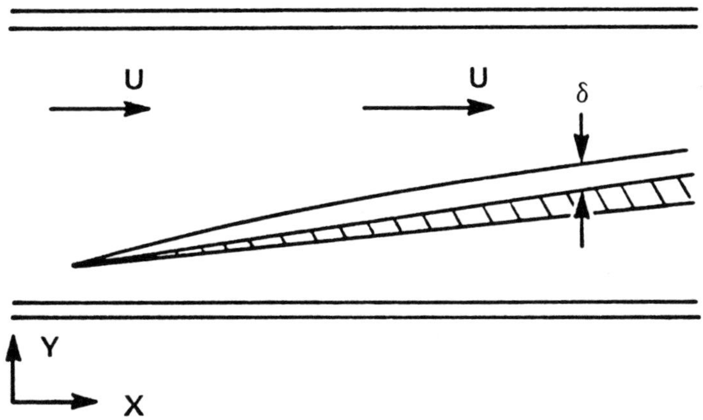

Figure 8: Reactor with tilted susceptor.

Turning our attention to surface phenomena rather than diffusion, we recognize that species will transit across the boundary layer and may be created or destroyed in this passage due to chemical reactions which will proceed at finite rates (homogeneous gas phase reactions). Upon impacting the surface, they may adsorb and then decompose, leaving a solid thin film. This will be a heterogeneous surface reaction which will have a characteristic chemical reaction rate. One way to describe this phenomena is in terms of a "mass transfer" coefficient. The mass flux can be expressed in terms of this coefficient, as follows:

$$J = \frac{k_D}{RT}(p_s - p_{eq}) \qquad (29)$$

where p_{eq} refers to the partial pressure of species i that would exist under equilibrium at the surface temperature, and k_D is the mass transfer coefficient.

For diffusion-controlled situations, Equation (28) shows that the driving force for mass flux to the surface is the pressure difference of species i between the main gas flow and the wall. For surface-controlled reactions, the driving force instead is the supersaturation at the surface. In other words, $p_s > p_{eq}$, and there are more gas molecules available at the surface than can be reacted there.

1.3.2 Effects of Gas Phase Kinetics

To properly describe chemical vapor deposition, one must develop a system of equations that encompasses all phenomena involved. This includes a proper representation of reactions in the gas phase, a suitable description of the surface kinetics, and the gas dynamics of a reacting gas mixture. Because the full governing equations are extremely complex and difficult to solve, most authors have examined only limited regimes. For example, we can ignore the gas dynamics completely and only study the kinetics of the gas phase reactions. Or, we could look at the kinetics of the heterogeneous surface reactions. If we also wish to ignore gas phase kinetics, we can study the thermodynamic description of the reaction.

Unfortunately, chemical vapor deposition is a field which is, basically, interdisciplinary. Essential understanding can be gained by including all of the phenomena involved. To do this in full generality would require the solution of many coupled, nonlinear, partial differential equations. Such a formulation is clearly beyond the scope of this text.

We will, therefore, choose to look at a particular simplified physical problem, but attempt to formulate the problem from first principles. This should lead to a problem, which although complex and involving nonlinear equations, can at least be described by ordinary differential equations.

Let us consider laminar flow through a two-dimensional channel, the so-called Poisseulle flow. The geometry is shown in Figure 9.

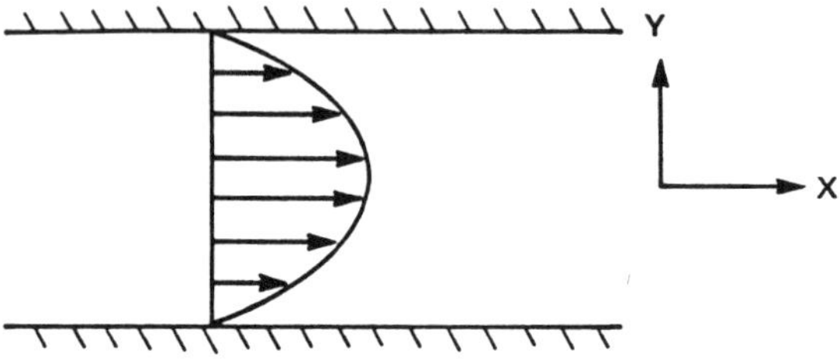

Figure 9: Channel flow.

As in the classical Poisseulle flow, the channel is assumed to be two-dimensional (nothing varies in the z direction) and doubly infinite in the x direction. We will impose a constant temperature, T, on each wall and assume that T will not vary with x. An axial pressure gradient will have to exist in order for there to be an axial flow, and we recognize that it should be constant so that p will vary linearly with x.

In a typical CVD reactor, mass flow is small so the pressure gradient will be small. Since p, T and density ρ are related by the equation of state, we can expect $\rho = \rho(x)$. However, the variation in density with x will be small, and it will be reasonable to neglect it.

As the reacting gas flows down the channel, it interacts with the channel walls, decomposes, and leaves a film on these walls. If the wall deposit is rapid and heavy, the reactants will deplete so that the gas composition will vary with x. Although this is a technologically important case, it requires a two-dimensional (partial differential equations) description. For the present problem, we will assume that depletion is slow enough for us to neglect, and gas composition will not be a function of x.

As in the classical Poisseulle flow, the y component of velocity will be zero, so that the overall mass continuity equation is identically satisfied. For a steady-state flow, we can write the simplified governing equations describing the velocity, temperature, and species conservation fields.

Momentum Conservation:

(30) $$\frac{d}{dy}\left[\mu \frac{du}{dy}\right] = \frac{dp}{dx}$$

where μ (gas viscosity) is a function of T and gas composition.

Energy Conservation:

(31) $$\frac{d}{dy}\left[k \frac{dT}{dy}\right] = \sum_{j=1}^{n} \rho Y_j V_{y_j} c_{p_j} \frac{dT}{dy} + \sum_{j=1}^{n} w_j c_{p_j} T \dot{\omega}_j$$

where
k = thermal conductivity (a function of T and gas composition)
Y_j = mass fraction of species j
c_{p_j} = specific heat at constant pressure of species j
$\mathring{\omega}_j$ = molar production rate of species j
w_j = molecular weight of species j
V_{y_j} = y component of diffusion velocity for the j species.

Species Mass Conservation:

(32) $$\frac{d}{dy}\left[\rho Y_j V_{y_j}\right] = w_j \dot{\omega}_j$$

and there is one such equation for each species j.

In order to complete specification of these equations, we have to express the diffusion velocity in terms of the species concentrations. We have,

(33) $$V_{y_j} = -\frac{D_j}{X_j}\frac{dX_j}{dy} + V_{c_y}$$

where V_{c_y} is a constant chosen to ensure the condition

$$\sum_{j=1}^{n} V_{y_j} Y_j = 0$$

obtained by summing Equation (32) over all species, the mole fraction X_j is related to the mass fraction, Y_j, by

$$\bar{w} Y_j = w_j X_j$$

The diffusion coefficient, D_j, refers to the diffusion of species j through the entire gas mixture. It can typically be evaluated approximately from the binary diffusion coefficients, which refer only to binary gas mixtures.[15] The latter can be calculated from rigorous kinetic theory.

Similarly, the viscosity and thermal conductivity can be evaluated approximately with the help of kinetic theory arguments.[15] Finally, we need an equation of state relating p, ρ, and T. Assuming we are dealing with a mixture of perfect gases, we have

$$p = \rho R T \tag{34}$$

where R is the mixture gas constant which is equal to $\mathcal{R}/\bar{w}$, with $\mathcal{R}$ the universal gas constant (1.987 cal/mole°K) and $\bar{w}$ the average molecular weight of the gas.

In order to solve these equations, we have to be able to evaluate $\dot{\omega}_j$, the species net production rate as a function of conditions and gas composition. If we assume only binary reactions and an Arrhenius temperature dependence for the forward rate coefficients of such reactions, then we can express $\dot{\omega}_j$ in a reasonably simple form.

First, let's choose the simple reaction of silane pyrolosis and solve our simplified equations for this case. Then, we have

$$SiH_4 \underset{k_r}{\overset{k_f}{\rightleftharpoons}} SiH_2 + H_2$$

so that there are three species to keep track of.

Then, if we refer to them as

$$SiH_4 = 1 \qquad SiH_2 = 2 \qquad H_2 = 3,$$

we have

$$\dot{\omega}_1 = - k_f [X_1] + k_r [X_2][X_3] = - \dot{\omega}_2 = - \dot{\omega}_3 \tag{35}$$

where k_f and k_r are the forward and reverse reaction rates of our one reaction equation, and $[X_1]$, and $[X_2]$ and $[X_3]$ are molar concentrations. As should be obvious, destruction of one SiH_4 molecule produces one SiH_2 and one H_2.

The forward rate coefficient is

$$k_f = A \exp [-E/RT] \quad , \tag{36}$$

where A and E are experimentally determined constants. The reverse rate coefficient is related to the forward one at equilibrium by

$$k_r = k_f/K_c \quad ,$$

where K_c is equilibrium constant in concentration units. Since we are dealing with a quasiequilibrium, we will use this to determine k_r. It is simpler to determine it from its pressure units form. The relationship between these forms is, for our case,

$$K_c = K_p \frac{p_{atm}}{T}, \tag{37}$$

where p_{atm} is atmospheric pressure, and K_p can be obtained from

$$K_p = \exp\left(\frac{\Delta S^o}{R} - \frac{\Delta H^o}{RT}\right) \tag{38}$$

where ΔS° is the change in entropy of the gases in our reaction in going from reactant to products under standard state conditions (atmospheric pressure). Then ΔH° is, similarly, the change in standard state enthalpy. The standard entropies, enthalpies and specific heats at constant pressure are all tabulated in the JANAF Table.[4] We can now express the species production rates as

$$\dot{\omega}_{SiH_4} = \left\{-\left[X_{SiH_4}\right] + \frac{RT[X_{SiH_2}]\,[X_{H_2}]}{p_{atm}\exp\left[\frac{\Delta S^o}{R} + \frac{\Delta H^o}{RT}\right]}\right\} A\exp[-E/RT] \tag{39}$$

and

$$\dot{\omega}_{SiH_2} = \dot{\omega}_{H_2} = -\dot{\omega}_{SiH_4}$$

or replacing species concentrations by species mass fractions, this becomes

$$\dot{\omega}_{SiH_4} = \left\{-\frac{Y_{SiH_4}}{W_{SiH_4}} + \frac{\bar{W}p\,Y_{SiH_2}\,Y_{H_2}}{W_{SiH_2}\,W_{H_2}\,p_{atm}\exp\left[\frac{\Delta S^o}{R} + \frac{\Delta H^o}{RT}\right]}\right\}\rho A\exp[-E/RT] \tag{40}$$

In order to proceed with calculations, C_p for each species and $\Delta H^\circ/RT$ plus $\Delta S^\circ/RT$ can be expressed as functions of temperature using the JANAF Tables.[4]

Finally, we have to define the proper boundary conditions for these equations. The boundary conditions for velocity and temperature are clear. They are:

$$y = 0; \quad u = 0 \quad T = TH$$
$$y = H; \quad u = 0 \quad T = TC.$$

The boundary conditions on species are not so simple. We have to determine Y_{SiH_4}, Y_{SiH_2}, and Y_{H_2} at $y = 0$ and H. Now, SiH_2 is an unsaturated molecule, so we assume that each molecule that strikes a surface reacts with unit probability. In that case, the proper boundary condition is

$$Y_{SiH_2} = 0 \text{ at } y = 0, H$$

where each SiH_2 molecule leaves one Si atom on the surface and one H_2 molecule leaves the surface.

The flux of SiH_4 molecules into a solid surface depends on whether they are destroyed at the surface or reflected. If reflected, the net flux is zero. If destroyed, the net flux is a maximum. For SiH_4, some are reflected and some are destroyed.

The fraction of silane molecules that adsorb and decompose upon collision with a solid surface can be estimated from experimental data.

Then the boundary condition on Y_{SiH_4} can be derived by equating the flux as calculated from continuum arguments to the flux, as computed from kinetic theory. The result is a mixed, nonlinear boundary condition involving Y_{SiH_4} and dY_{SiH_4}. From this, we can evaluate the rate at which a silicon film will grow on the hot wall.

The boundary condition on Y_{H_2} can be determined if we remember that each SiH_2 and each SiH_4 molecule releases H_2 as it decomposes on the surface. Then we can write

$$\frac{V_{H_2} Y_{H_2}}{W_{H_2}} = \frac{V_{SiH_4} Y_{SiH_2}}{W_{SiH_4}} + \frac{2V_{SiH_4} Y_{SiH_4}}{W_{SiH_4}} \tag{41}$$

evaluated at the wall.

Finally, we require expressions for μ, k, and D_j as functions of T and Y_j's before we can solve our equations. As noted earlier, they can be derived from kinetic theory, and an explanation of how they are developed is available.[15]

Equations (30), (31), and (32) are all highly nonlinear differential equations, so we will solve them by replacing derivatives with finite differences and use a high-speed digital computer to solve the resulting difference equations.

Before discussing solution techniques, it is interesting to make the following observations:

(1) The momentum equation depends only on T through the temperature dependency of μ.

(2) The energy equation requires a knowledge of the Y's, but is independent of u.

(3) The species conservation equations depend on T, but are also independent of u.

Therefore, we can solve the energy and species equations to obtain values for the Y's and T, and then use these to calculate u.

The boundary conditions for the solution are:

$$u(o) = u(L) = 0$$

$$T(o) = TH, \quad T(L) = TC$$

and the conditions on the Y's discussed earlier.

The momentum and energy equations are solved using a point-by-point iteration scheme. Derivatives are first replaced by finite differencies. A typical point is shown below

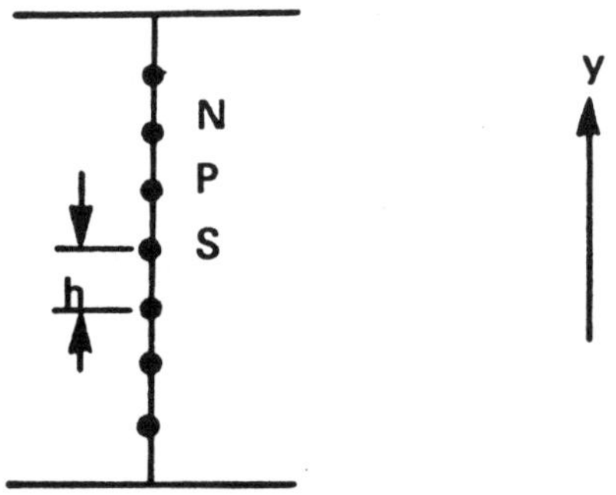

and we write, for any functions, ϕ and ψ

$$\frac{d\phi}{dy} = \frac{\phi_N - \phi_S}{2h}$$

and

$$\frac{d}{dy}\left[\psi \frac{d\phi}{dy}\right] = \frac{1}{2h^2}\left[(\psi_N + \psi_P)(\phi_N - \phi_P) - (\psi_P - \psi_S)(\phi_P - \phi_S)\right]$$

Then, Equations (30) and (31) are written as

Momentum:

$$\frac{1}{2h^2}\left[(\mu_N + \mu_P)(u_N - u_P) - (\mu_P + \mu_S)(u_P - u_s)\right] = \frac{dp}{dx} \tag{42}$$

Energy:

$$\frac{1}{2h^2}\left[(k_N + k_P)(T_N - T_P) - (k_P + k_S)(T_P - T_S)\right. =$$

$$\sum_{j=1}^{n} \left\{ \rho_P V_{j_P} (V_{yj})_P (c_{P_j})_P \frac{[T_N - T_S]}{2h} + \dot{\omega}_{j_P} w_j (c_{P_j})_P T_P \right. \tag{43}$$

In the energy equation, we can replace $(V_{y_j})_P$ from Equation (33) so that Equation (45) can be rewritten as:

$$\frac{1}{2h^2}\left[(k_P + k_N)(T_N - T_P) - (k_P + k_S)(T_P - T_S)\right] =$$

(44)
$$\sum_{j=1}^{n} \rho_P\, Y_{j_P} \left\{ - \frac{D_{j_P}}{Y_{j_P}} \frac{\left[X_{j_N} - X_{j_S}\right]}{2h} + \sum_{k=1}^{n} \frac{D_{k_P} Y_{k_P}}{Y_{k_P}} \frac{\left[X_{k_N} - X_{k_S}\right]}{2h} \right\}$$

$$\left(c_{P_j}\right)_P \frac{[T_N - T_S]}{2h} + \sum_{j=1}^{n} \dot{\omega}_{j_P}\, w_j \left(c_{P_j}\right)_P T_P$$

or, using $\bar{w}Y_j = w_jX_j$ and $p = \dfrac{\rho RT}{\bar{w}}$, we get

$$\frac{T_P}{2h^2}\left[(k_N + k_P)(T_N - T_P) - (k_P + k_S)(T_P - T_S)\right] =$$

$$\frac{(T_N - T_S)}{4h^2}\, \frac{p\,\bar{w}_P}{R} \sum_{j=1}^{n} \left(c_{P_j}\right)_P Y_{j_P} \left\{ - \frac{D_{j_P}\left[\bar{w}_N\, Y_{j_N} - \bar{w}_S\, Y_{j_S}\right]}{\bar{w}_P\, Y_{j_P}} \right.$$

$$\left. + \sum_{k=1}^{n} \frac{D_{k_P}}{w_P} \left(\bar{w}_N\, Y_{k_N} - \bar{w}_S\, Y_{k_S}\right) \right\} + \sum_{j=1}^{n} \dot{\omega}_{j_P}\, w_j \left(c_{P_j}\right)_P T_P^2$$

For a small degree of dissociation, we assume $\bar{w}_n = \bar{w}_s = \bar{w}_p$, and Equation (46) can be simplified as

$$\left[2k_P + k_N + k_S + 2h^2 \sum_{j=1}^{n} \dot{\omega}_{j_P}\, w_j \left(c_{P_j}\right)_P\right] T_P^2$$

(45)
$$- \left[(k_N + k_P)T_N + (k_P + k_S)T_S\right] T_P$$

$$- \frac{(T_N - T_S)}{2}\, \frac{p\bar{w}_P}{R} \sum_{j=1}^{n} \left(c_{P_j}\right)_P Y_{j_P} \left\{ - \frac{D_{j_P}\left(Y_{j_N} - Y_{j_S}\right)}{Y_{j_P}} + \sum_{k=1}^{n} D_{k_P}\left(Y_{k_N} - Y_{k_S}\right) \right\} = 0$$

At a typical grid point, we assume we know T_N and T_S and wish to solve for T_P. If the iteration is proceeding upward (y positive), then T_S for the first interior point is known from the boundary condition and T_N is known from the initial guess. The thermal conductivity, k, the net production rates, $\dot{\omega}_j$, and the diffusion coefficients, D_j, are calculated from the initial guess for T and the assumed known solution for the Y's. We then solve the quadratic equation for T_p at the first interior point. Next, the following point is considered and T_S for it is the just-calculated T_p from the first point. In this way, we calculate T at each point up to the upper boundary. Then, with a new estimate of T available, we recalculate k, $\dot{\omega}_j$, and D_j and repeat the procedure.

Next, we have to solve for the Y_j's from the species continuity equations, Equation (32). Unfortunately, these equations cannot be integrated by a similar simple point iteration scheme as they are mathematically "stiff"[16] and iterative approaches are unstable. To solve these simultaneous equations, we turn to a perturbation analysis developed by Newman[17] where the equations are linearized about an initial guess, and the resulting linear equations are solved numerically. The solution is then used as the next guess, and the linear equations are resolved. The procedure is repeated until the solution no longer changes.

If there are n species, we have n simultaneous linear ordinary differential equations, which can be solved by well-known techniques.

Typically, 7-10 iterations are needed to achieve convergence if an adequate number of grid points have been chosen. For problems involving chemical kinetics, this can be a large number, which leads to a lengthy calculation. For some of the cases we calculated, it was necessary to use 3000 grid points over a 3-cm channel height to secure convergence.

Once the Y_j's have been calculated, we can recalculate the temperatures across the channel. Then, the corrected temperatures can be used to generate a new set of Y_j's. When the T and Y_j arrays no longer change, the flow field (u) can be calculated directly, since we can then calculate u as a function of channel height.

It should be noted that although it may use a large number of points to solve for the Y_j's, a large number is not necessary to obtain accurate representations of T and u.

For these calculations, attention was limited to a temperature range of 950° to 1350°K, a pressure range of 300 mTorr to 7.6 Torr, and SiH_4 mass fractions of 15 to 30% in H_2. Under these conditions, the mass fraction of SiH_2 formed was at most on the order of 10^{-3}, so that the influence of the Y_j's on the temperature distribution was small. To a good degree of approximation, the temperature was calculated to be linear. By the same token, the flow field which was easily calculated did not demonstrate any unique behavior.

Most of the effort was spent trying to integrate the three simultaneous Y equations. The Y distributions across the channel for a typical condition are shown in Figure 10. The Y_{SiH_2} exhibits a peak near the hot wall and is a fairly full profile. This can be attributed to the high diffusion coefficient at these pressures, which allows the SiH_2 to readily diffuse toward the cold wall. Deposition on the cold wall is many times smaller than on the hot wall, as evidenced by the smaller value of dY_{SiH_2}/dy there.

Deposition rates as a function of hot wall temperature are presented in Figure 11 with pressure as the parameter, and Figure 12 with mass fraction as the parameter. For the temperature range studied here, there is no evidence of a reduction in the rate of increase in deposition rate as the temperature is increased. Variations with pressure and mass fraction are as would be expected.

It is interesting to compare the present results with data obtained in a hot wall furnace[18] tube, even though the present calculations are for one hot and one cold wall and a different physical arrangement. For one case, deposition rate was measured at p = 532 mTorr, Y = 2.3% and T = 898°K. Without running the exact case numerically, we can estimate from Figures 11 and 12 a calculated value of 2.5 Å/min compared to a measured value of 4 Å/min.

Having the numerically calculated value on the same order of magnitude as experimentally-measured values lends credibility to the model being used—especially since the model has been developed from first principles and involves no adjustable parameters.

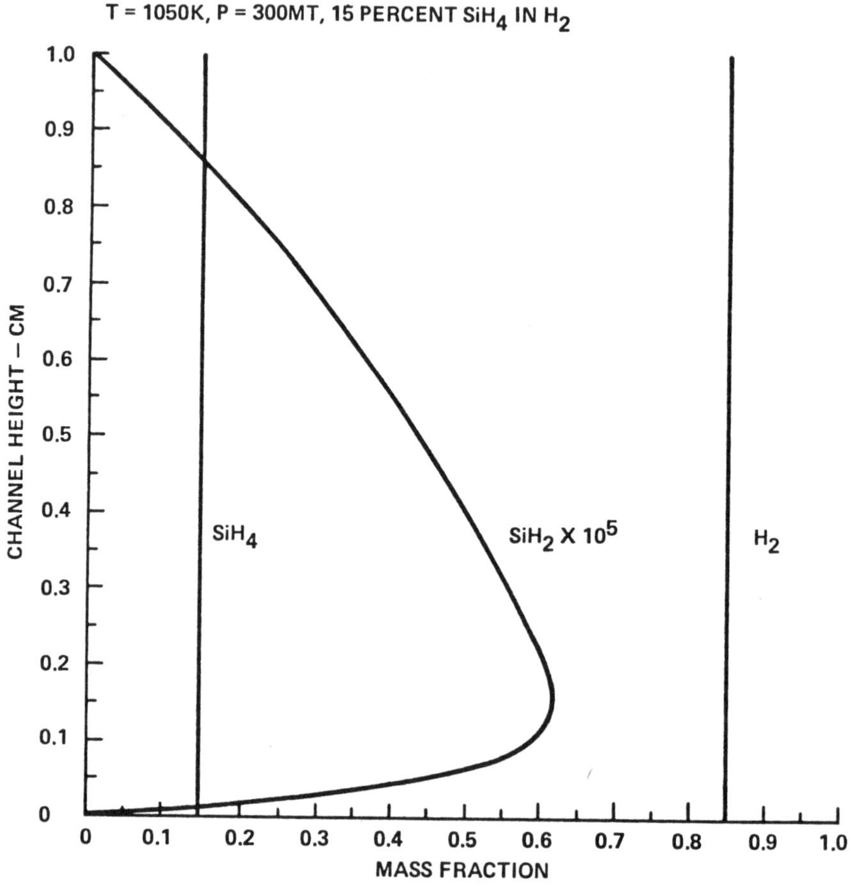

Figure 10: Mass fraction distribution of three species in SiH_4 pyrolysis into SiH_2 and H_2.

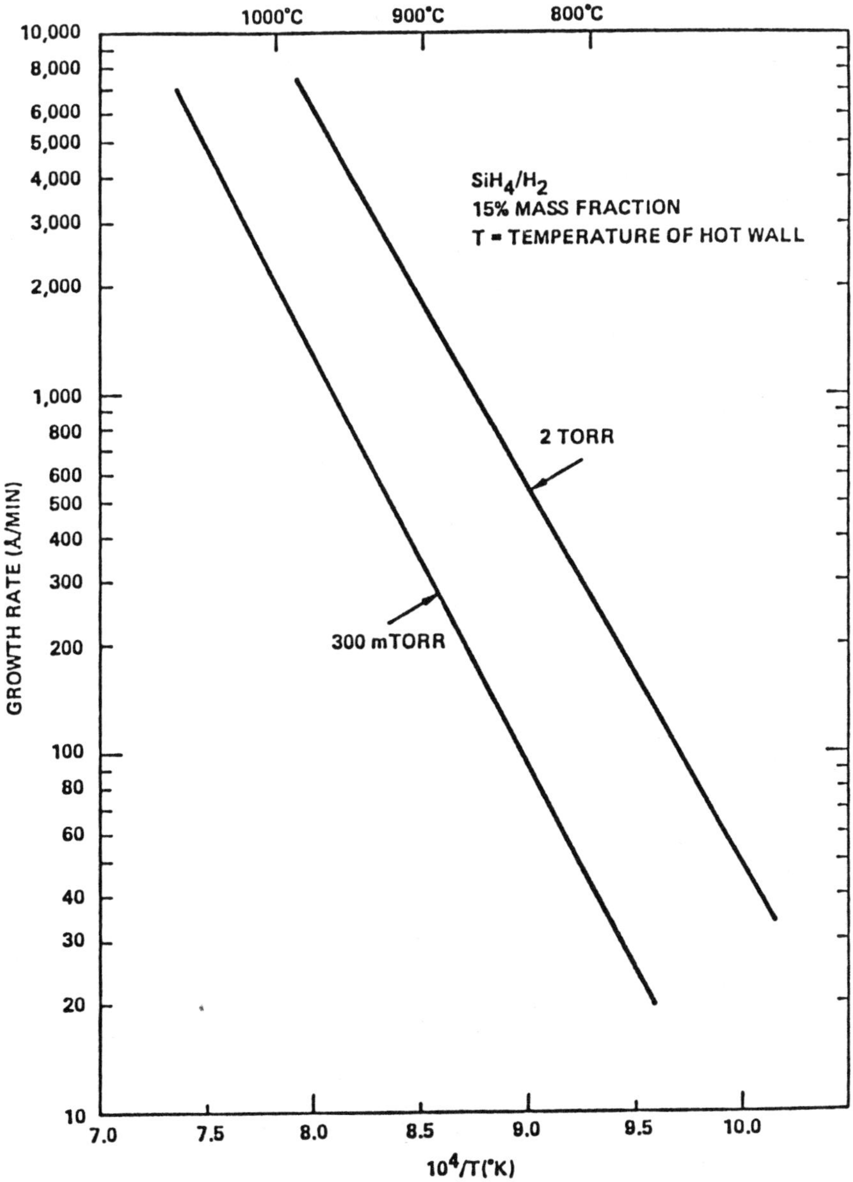

Figure 11: Silicon film growth rate as a function of temperature for different pressures.

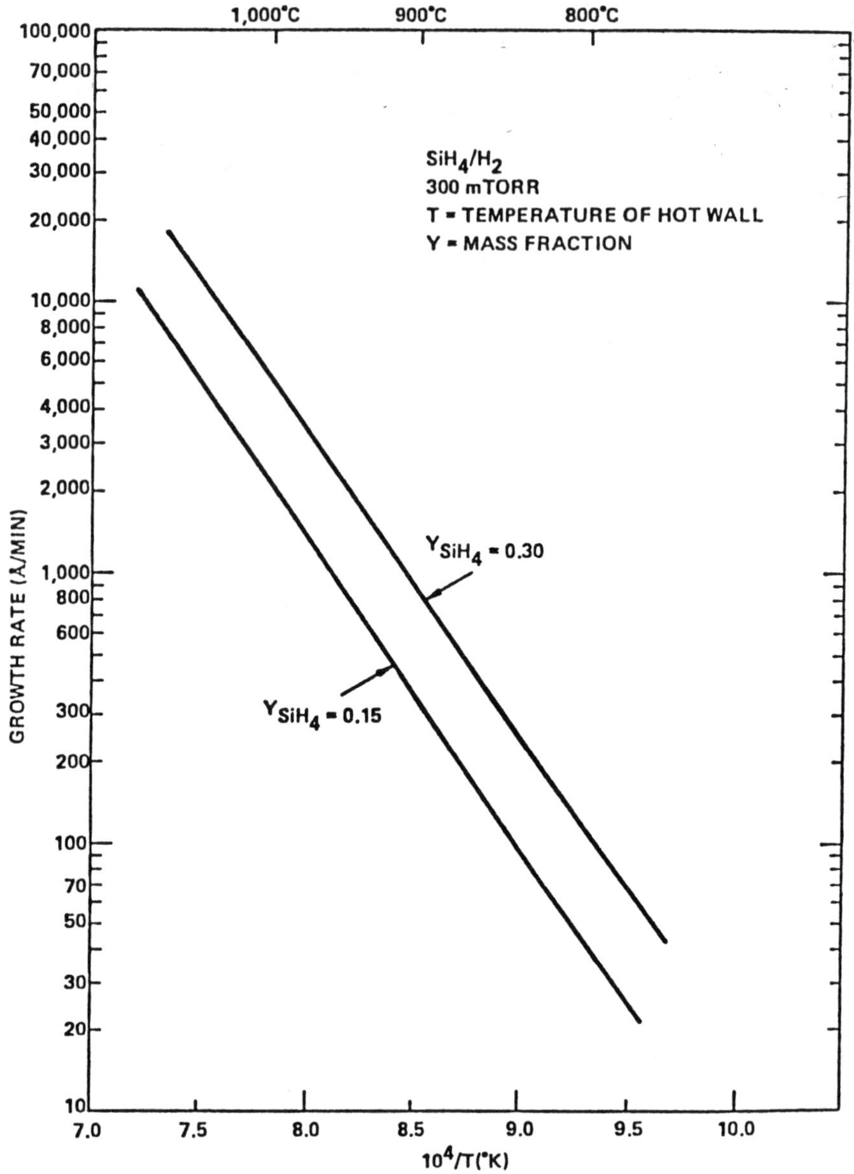

Figure 12: Silicon film growth rate as function of temperature for different initial mass fractions.

1.4 FILM MORPHOLOGY

There is a substantial literature describing the structure of thin films grown by evaporation, sputtering, and CVD for a variety of applications.[19,20] Such films can initially be grown as amorphous, polycrystalline, or single-crystal materials, depending on conditions during growth. Amorphous films can be converted to polycrystalline by annealing at high temperatures. The physical nature of the thin film material will play an important role in determining film properties, such as electrical resistivity, stress, adhesion, among others. Also, the surface roughness may be important in determining the ease with which optical lithography can be used in integrated circuit manufacture.

Films grown by chemical vapor deposition are similar to the films described above, with one exception. The CVD process allows for the possibility of gas phase particle nucleation, and the incorporation of such particles in a growing surface can contribute to a surface roughness. This is one reason that atmospheric CVD is being used less and less, as compared to low-pressure CVD where such gas phase nucleation is less likely.

In general, dielectric films such as SiO_2 and Si_3N_4 grow in CVD processes as amorphous films. Metallic films generally grow as polycrystalline with a typical columnar structure. Figure 13a shows a low-pressure CVD polysilicon film as deposited with such a structure.[19] Annealing of metallic films can alter this columnar structure, again depending on conditions. For undoped polysilicon films, annealing (1 hour, 950°C) leaves the film structure unchanged. However, when the polysilicon film is phosphorus doped, annealing substantially alters the grain structure, as shown in Figure 13b.

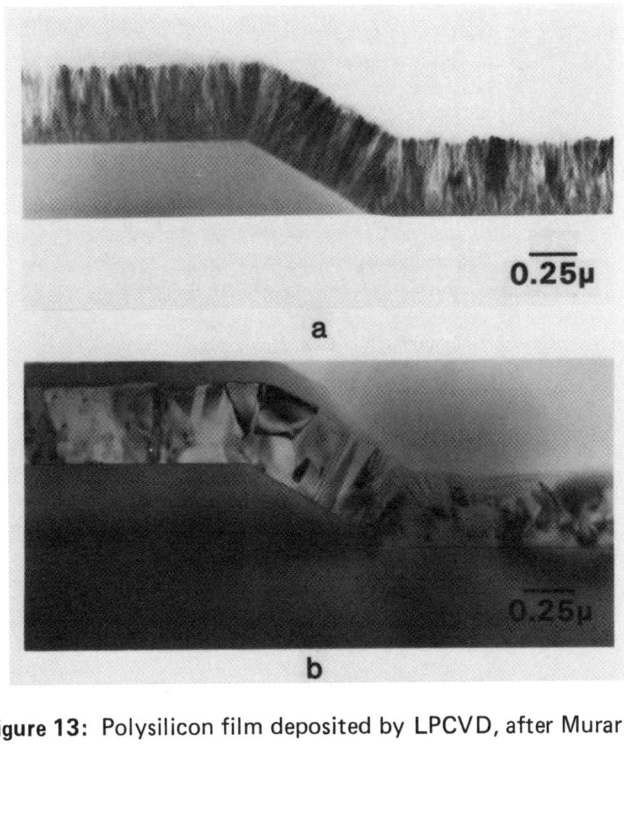

Figure 13: Polysilicon film deposited by LPCVD, after Murarka.[19]

CVD tantalum silicide and aluminum[21] go down as polycrystalline films, as shown by the TEMs in Figures 14 and 15. In general, any of the metallic films will be polycrystalline. The crystal size, however, will depend on processing history and/or heat treatment. As noted earlier, surface roughness may occur when these films are deposited. In particular, CVD aluminum grown in a hot tube[21] tends to exhibit a rough surface, as shown in an SEM in Figure 16.

Figure 14: SEM of CVD $TaSi_2$ on oxide.

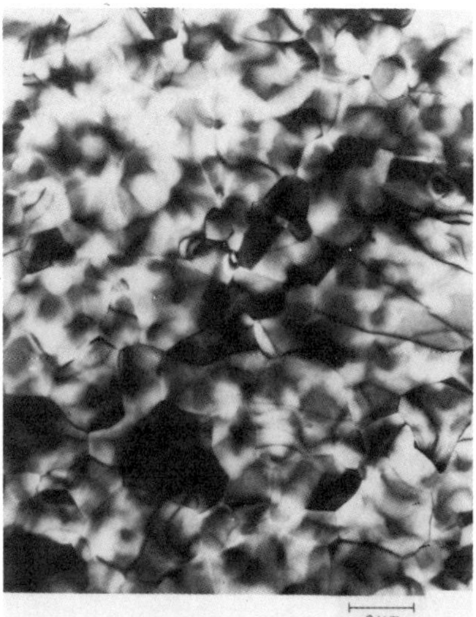

Figure 15: TEM of aluminum thin film deposited in LPCVD hot tube furnace. After Levy.[21] (Reprinted by permission of the publisher, The Electrochemical Society, Inc.)

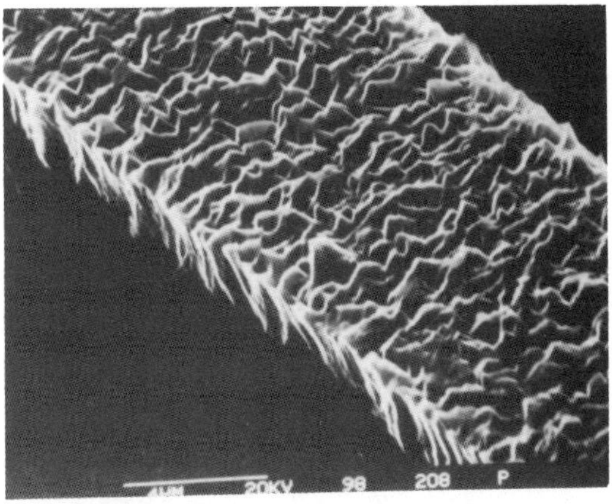

Figure 16: SEM showing a rough surface of LPCVD aluminum grown in a hot tube furnace. After Levy.[21]

1.5 LABORATORY THERMAL CVD REACTORS

In this section we will review the various types of CVD reactors scientists and engineers have used for the development of thermal CVD processes. This will be distinct from the commercial reactors used for production which will be covered in a later chapter. A similar review of reactors for development of plasma-enhanced CVD processes will be made at the end of the next chapter. We will cover the so-called cold wall systems for either single or multiple wafers first, followed by a discussion of continuous belt systems. Finally, we will review the hot wall reactor approach.

1.5.1 Cold Wall Systems—Single Wafer

When we speak of a cold wall CVD reactor, we refer to a continuous flow system where the wafer is kept at the required high temperature, but all other surfaces bounding on the reacting gases are cold. The objective here is to cause the desired reaction only on the hot wafer and keep all other surfaces free of deposits. In practice this is a goal that can only be partially attained. Although reactions will proceed more slowly on colder surfaces, they will proceed—and films will build up. At the same time the films that form on the colder surfaces may be more porous than the normal film and may spall off more easily. All of which says that in spite of our best efforts, cold walled reactors may have their cold walls an undesirable source of particulates which may end up on the hot substrate. The occurence of such particulates can be minimized by frequent cleaning of the chamber walls to remove deposits.

1.5.1.1 Tube Reactor, Parallel Flow: A sketch of this reactor configuration is shown in Figure 17. Here the reactant gas or gas mixture flows axially down a tube (circular or rectangular cross section) over a heated susceptor aligned parallel to the flow. The wafer to be coated is placed on this susceptor. The tube is most often fused quartz. Techniques for heating the susceptor and not the tube walls will be discussed later. Operation can be at atmospheric or at low pressures depending on the process being developed. Low pressure operation will, of course, require vacuum pumping capability.

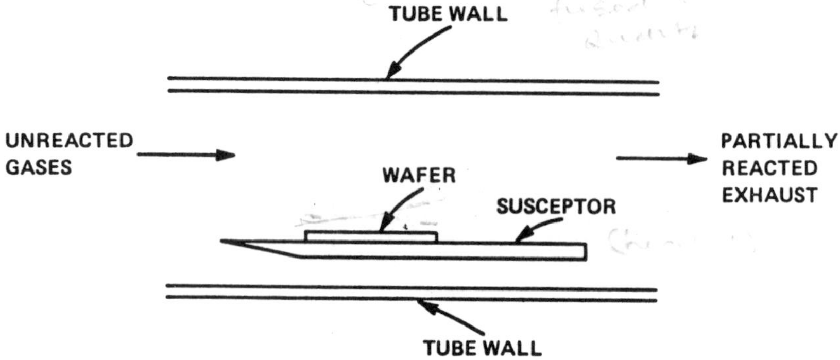

Figure 17: Tube reactor, parallel flow.

1.5.1.2 Tube Reactor, Normal Flow: An alternative arrangement for such a reactor would be to align it vertically, and place the wafer on a pedestal normal to the flow direction. Such a system is shown in Figure 18.

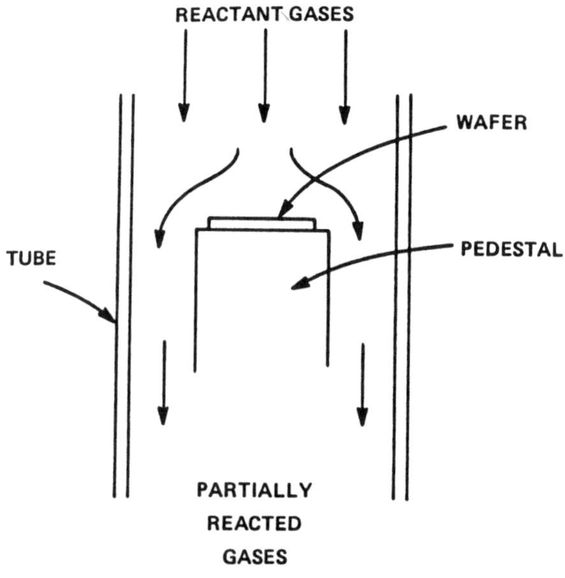

Figure 18: Tube reactor, normal flow.

The difference in these two configurations is in the flow pattern developed over the wafer. In Figure 17 the flow will develop as a classical boundary layer which will be thinner at the front and thicker towards the trailing edge. To whatever extent the deposition is diffusion controlled, it will be influenced by boundary layer thickness, so deposition rates may be higher near the front of the wafer than in the rear. For Figure 18 the flow pattern will be quite different and whatever nonuniformities occur will at least be axisymmetric.

1.5.1.3 Heating Systems: Over the years, three methods of heating the susceptor but not the tube walls have been used. For one, the susceptor can be made of an electrically conducting material (i.e., graphite) and used as a resistor with an electrical power supply. Joule heating will then heat it readily.

Alternately, one could heat the conducting susceptor inductively with a conducting coil placed around the tube, as shown in Figure 19. The coil would be activated by an AC power supply.

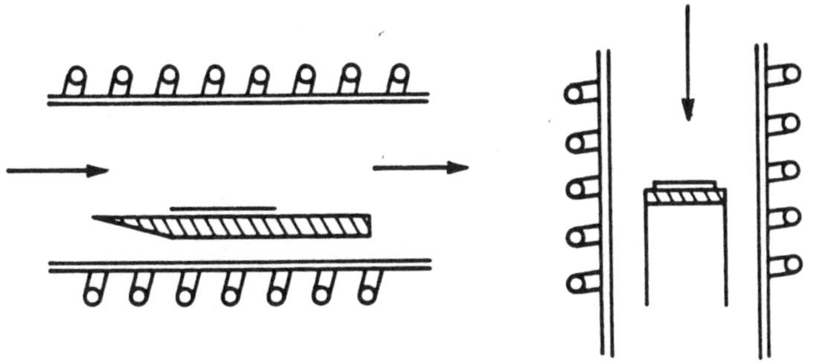

Figure 19: Inductive heating of tube reactors.

Finally, heating can be done by irradiating the susceptor with high intensity lamps that will transmit readily through the fused quartz.

Direct heating of a resistive substrate requires high temperature electrical connections within the reactor proper. For this reason, inductive heating seems to be preferred. Optical heating has one advantage over either of the first two schemes. When the susceptor is heated electrically, the wafer sits on a heated surface and its bottom surface can be somewhat hotter than its top surface. For certain high temperature processes, this small difference can be enough to make the wafer warp, which can cause disturbances in the flow or in extreme cases damage (i.e., slip) to the wafer. With optical heating, one can heat the wafer from above while it is inductively heated from below, or from both above and below, so that the wafer remains in a uniform temperature environment.

1.5.2 Cold Wall Systems—Multiple Wafers

Obviously, it would be desirable to process more than one wafer at a time in a CVD reactor, since wafer throughput (i.e., wafers/hour) could be an important factor in determining how commercial a process will be. Multiple wafer reactors can all trace their roots to some variant of the two simple systems just described.

1.5.2.1 Tube Reactor: When initial attempts were made to extend the susceptor of Figure 17 in the flow direction, it was discovered that deposits on the last wafers were thinner than deposits on the initial ones. As discussed earlier, the explanation proposed[14] was that deposition was diffusion controlled so that the thicker downstream boundary layer limited the diffusion of reactant species to the downstream wafer. An obvious solution is to tilt the susceptor up towards the rear (see Figure 8). This would have the effect of thinning the downstream boundary layer which should correct this problem. Experimental data[14] have verified this effect as shown in Figure 20.

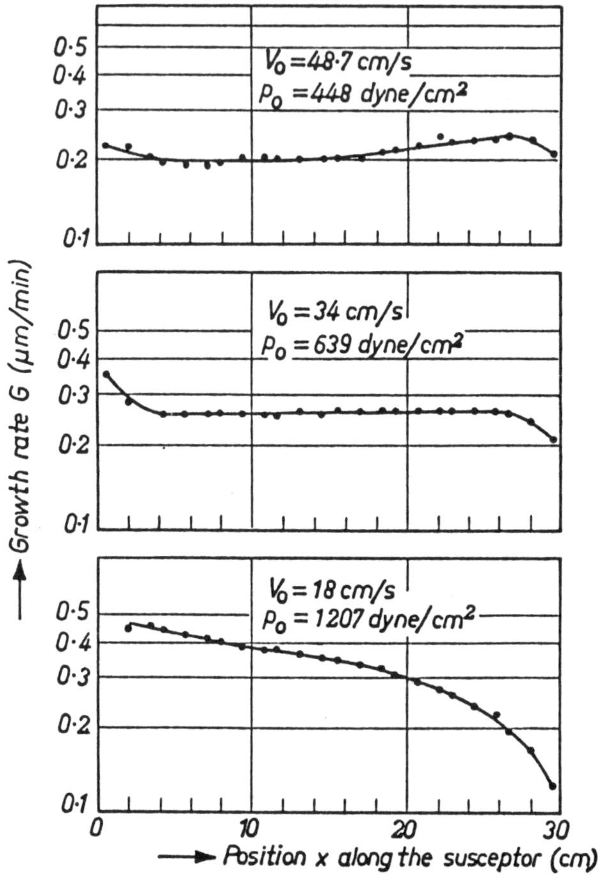

Figure 20: Deposition along length of tilted susceptor.[14] Reprinted by permission of the publisher, The Electrochemical Society, Inc.

1.5.2.2 Bell Jar Reactor, Barrel Susceptor: If we wanted to place even more wafers in a reactor, one can go to a bell jar configuration and extend the tilted susceptor tube concept just described. Consider a reactor within a bell jar arranged as shown in Figure 21.

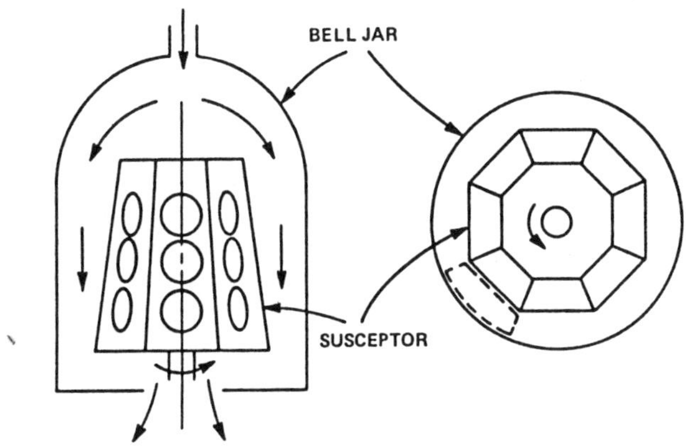

Figure 21: Bell jar reactor, barrel susceptor, axial flow.

In this configuration, reactant gases enter at the top of the bell jar and flow between the bell jar and susceptor before exiting at the bottom. The susceptor is multifaceted, the number of facets being determined by wafer size and bell jar diameter. It is tilted to enable wafers to sit in pockets and be held in place by gravity. If we look at one sector of this reactor (delineated by the dotted lines) we see that we have simply replicated a tube reactor with a tilted susceptor several times around the bell jar periphery. The narrowing of the flow passage towards the exit again serves to thin the exiting boundary layer. The entire susceptor can be rotated to minimize any tangential nonuniformities.

With this reactor design, the susceptor can be heated inductively by wrapping a coil around the bell jar, or radiantly from outside the bell jar.

1.5.2.3 Bell Jar Reactor, Barrel Susceptor, Radial Flow: In much the same way that we extended the tube reactor with parallel flow to a bell jar geometry, we can do the same with the tube reactor with normal flow. Consider Figure 22. Since the flow is entering radially from the outside, one way to heat the susceptor is with high powered lamps within the central cavity. Again, the susceptor could be rotated to improve uniformity.

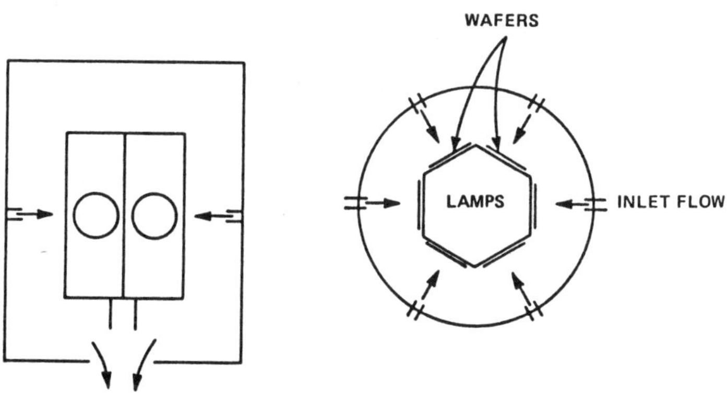

Figure 22: Bell jar reactor, barrel susceptor, radial flow.

1.5.2.4 Pancake Reactor: The tube reactor with normal flow leads to another type of bell jar reactor referred to as the pancake reactor. It is shown in Figure 23. Here the susceptor is a disc placed horizontally and heated by induction by coils placed below it. The reactive gas flow could be introduced from above, but the favored approach is to introduce it from below at the center of the susceptor disc. Gas exhaust is at the periphery between the disc and the bell jar.

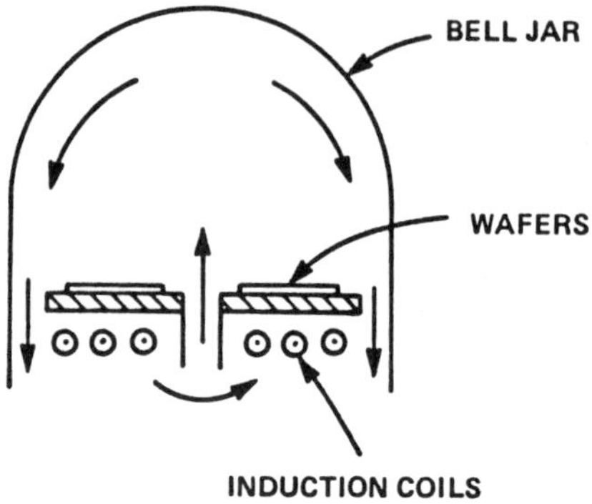

Figure 23: Pancake reactor.

1.5.3 Cold Wall Systems—Continuous Belt

These systems are similar to the tube reactor with normal flow except that wafers are passed horizontally under the reactant gas flow while being heated from below. One version uses a premixed gas stream and flow distributor long enough to deposit on many wafers simultaneously. A sketch of such a system is shown in Figure 24. Wafers are heated from below by resistance heaters radiating to the wafer carriers. Wafers enter on the left and leave on the right after passing through nitrogen purge curtains which protect the operators from any toxic gases used as reactants.

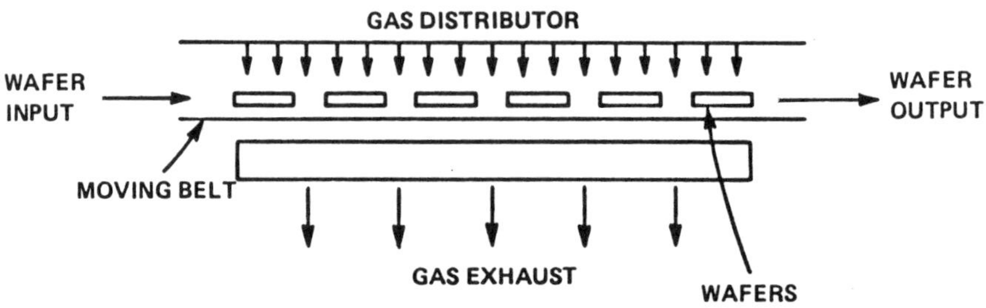

Figure 24: Continuous CVD reactor—premixed gas flow.

The primary advantage of these continuous systems is the high wafer throughput. Also they are run at atmospheric pressure so the expense of a vacuum pumping system can be avoided. The disadvantage is that any deposits on the cooled distributor plate end up as particles on the wafers.

The second approach is to use separate reactor streams that mix very close to the moving wafers. In Figure 25 we show a typical arrangement.

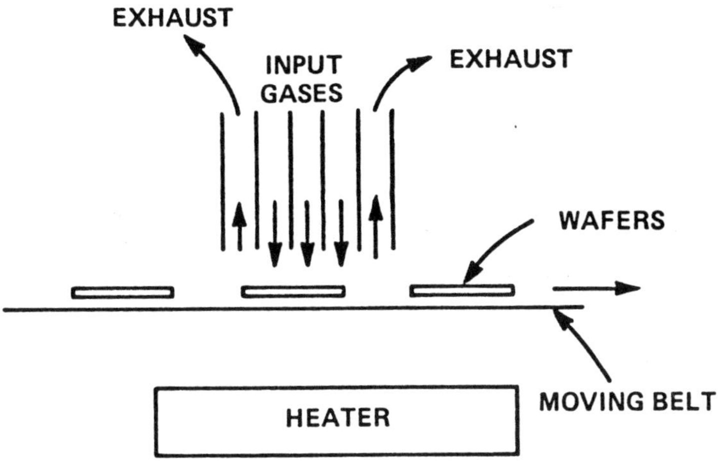

Figure 25: Continuous CVD reactor-separate gas flows.

1.5.4 Hot Wall Systems

One of the problems with cold wall systems is the difficulty in maintaining a very uniform temperature on the wafers. Such concern can be eliminated if the entire reactor chamber is placed within a furnace maintained at a very uniform temperature. An ideal candidate for such a furnace is the standard diffusion tube furnace already in wide use for integrated circuit fabrication. If in addition, wafers could be loaded vertically as in a diffusion furnace, the reactor throughput could be substantial.

The problem of assuring uniform depositions on many wafers closely spaced in a long uniform tube was solved when operation of the reactor at low pressure was considered.[22] Normally, in an atmospheric pressure cold wall CVD system, the reactant gas is heavily diluted in N_2 in order to reduce gas phase nucleation. At the pressures used for low pressure CVD (0.5-1.0 Torr), this is less of a problem so less diluent is needed. The net effect then is that deposition rates only fall by a factor of five. However, as many as 100 wafers can be processed in such a reactor at one time (see Figure 26), and this more than compensates for the lower deposition rate. In addition, due to the low pressure, diffusion occurs at high rates and the deposition tends to be controlled by the surface temperature. Given the very uniform temperatures available in a diffusion furnace, the deposition uniformity tends to be excellent in such a system.

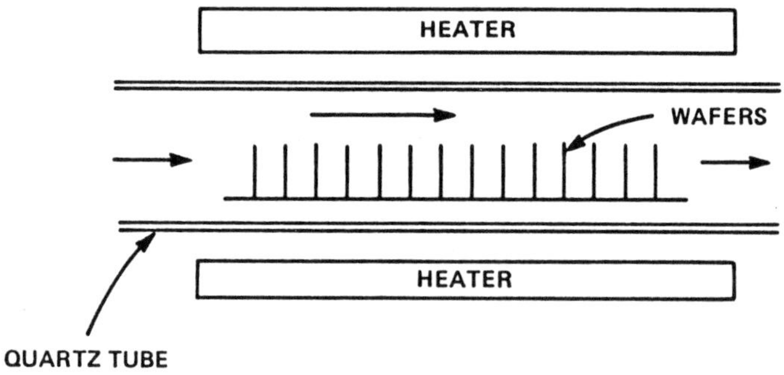

Figure 26: Low pressure hot wall CVD reactor.

REFERENCES

1. Glasstone, S., *Textbook of Physical Chemistry,* Princeton, NJ: D. Van Nostrand (1946).
2. Zeleznik, F.J., and Gordon, S., Calculation of complex chemical equilibria. *Industrial and Chemical Engineering* 60: 27 (1968).
3. Ban, V.S., and Gilbert, S.L., Chemical processes in vapor deposition of silicon. *JECS* 122: 1382 (1975).
4. "JANAF Thermochemical Tables." Dow Chemical Corp. (1965).
5. Kern, W. and Ban, V.S., in *Thin Film Processes* (J.L. Vossen and W. Kern, eds.) p. 257. Academic Press, NY (1978).
6. White, W.B., Johnson, S.M., and Dantzig, G.B., Chemical equilibrium in complex mixtures. *J. Chem. Phys.* 28: 751 (1958).
7. Eriksson, G., Thermodynamic studies of high temperature equilibria. *ACTA Chem. Scand.* 25: 2651 (1971).
8. Cruise, D.R., Notes on the rapid computation of chemical equilibria. *J. Phys. Chem.* 68: 3797 (1964).
9. Wan, C.F., and Spear, K.E., in: *Proc. Sixth International Conference on Chemical Vapor Deposition* (L.F. Donaghey, P. Rai-Choudhury, R.N. Tauber, eds.), p. 47, Electrochemical Society, Princeton, NJ (1977). These figures were originally presented at the Fall 1977 Meeting of the Electrochemical Society, Inc. held in Atlanta, Georgia.
10. Besmann, T.M., and Spear, K.E., Analysis of the chemical vapor deposition of titanium diboride. *JECS* 124:786 (1977).
11. Hunt, L.P., and Sirtl, E., A thorough thermodynamic evaluation of the silicon-hydrogen-chlorine system. *JECS* 119:1741 (1972).
12. Coltrin, M.E., Kee, R.J., and Miller, J.A., A mathematical model of the coupled fluid mechanics and chemical kinetics in a chemical vapor deposition reactor. *JECS* 131:425 (1984).
13. Sherman, A., Numerical modeling of CVD reactors including gas phase kinetics. To be published.

14. Eversteijn, F.C., Severin, P.J.W., v.d. Brekel, C.H.J., and Peek, H.L., A stagnant layer model for the epitaxial growth of silicon from silane in a horizontal reactor. *JECS* 117:925 (1970).
15. Kee, R.J., Warnatz, J., and Miller, J.A., A FORTRAN computer code package for the evaluation of gas-phase viscosities, conductivities, and diffusion coefficients. Sandia Report SAND83-8209 (March 1983).
16. Zucros, M.J., and Hoffman, J.D., *Gas Dynamics, Vol. II,* New York, J. Wiley & Sons (1977).
17. Newman, J.S., *Electrochemical Systems,* New Jersey: Prentice-Hall (1983).
18. Claassen, W.A.P., Bloem, J., Valkenburg, W.G.J., and Van den Brehel, C.H.J., The deposition of silicon from silane in a low-pressure hot-wall system. *J. Crys. Growth* 57:259 (1982).
19. Murarka, S.P., Interactions in metallization systems for integrated circuits, *J. Vac. Sci. Technol.* B2:693 (1984).
20. Bunshah, R.F. (editor), *Deposition Technologies for Films and Coatings,* Park Ridge, NJ, Noyes Publications (1982).
21. Levy, R.A., Green, M.L., and Gallagher, P.K., Characterization of LPCVD aluminum for VLSI processing. *JECS* 131:2175 (1984).
22. Rosler, R.S., Low pressure CVD production processes for poly, nitride, and oxide. *Solid State Techn.* 20(4):63 (1977).

2

Fundamentals of Plasma-Assisted CVD

2.1 INTRODUCTION

As discussed in the previous chapter, CVD of thin films depends on a heterogeneous surface reaction between a reacting gas and a hot surface. In many applications, the temperatures required to obtain commercially realistic deposition rates are prohibitive. For example, depositing titanium nitride on tool steel has to be done at temperatures above the softening temperature of the tool steel. This necessitates hardening the tool steel after the CVD deposition, making it difficult to maintain critical dimensions. A similar difficulty occurs if silicon nitride is to be deposited on integrated circuits as a final passivation layer. Since the final metallization is aluminum, which melts at about 600°C, and the temperature necessary to deposit Silicon Nitride thermally is 800° to 900°C, this clearly cannot work, although such a film would be an excellent barrier film.

One solution to such difficulties is to use a glow discharge in the reacting gas to create a higher than equilibrium quantity of free radicals.[1] The flux of such free radicals to a solid surface at low temperatures is then sufficient to permit acceptable deposition rates. For example, plasma-enhanced CVD of silicon nitride has been a commercial process in the integrated circuit industry[2] for some years. Good quality films are prepared at 350°C. Other films have also been deposited by the same technique. These include amorphous silicon,[3] iron oxide,[4] epitaxial silicon,[5] as well as refractory metal and metal silicides,[6] among others. Silicon Dioxide films have also been produced by this technique[7] in spite of the fact that Silicon Dioxide can be deposited thermally at about 400°C. This points out the fact that there are often reasons other than low temperature deposition for using plasma-enhanced CVD. For silicon dioxide, the interest is in varying the film stoichiometry, which cannot be done thermally. For amorphous silicon used for solar cells, a substantial quantity of hydrogen is incorporated in the film when a plasma is used. The hydrogen incorporated enables the amorphous silicon to behave as a semiconductor.[3]

Commercial plasma-enhanced CVD reactors all operate with RF glow discharges, creating a low pressure plasma. The reacting gases are ionized and dissociated by electron impact, and in the bulk of the plasma, there are equal numbers of electrons and ions. The electron temperature can be on the order of 20,000°K or higher, while the gas temperature remains near room temperature, depending on the pressure at which the discharge is operated. Along electrode surfaces, a thin sheath forms where electrical charge neutrality is no longer maintained. The behavior of the electrons, ions, and atoms as they traverse the sheath on the way from the plasma to a surface, often determines the nature of the resulting film.

In addition to plasma-enhanced CVD reactors that we will study in this text, there is another important commercial application for low-pressure plasmas in "dry" etching.[8] Here, a plasma converts an inert gas (i.e., from $-CF_4$) to a mixture including free fluorine, which can then etch Si, SiO_2, etc. There are two advantages. One, the starting gas is inert and, two, anisotropic etching is possible due to ion bombardment.

In the balance of this chapter, we will discuss plasmas in more detail to establish a basis for a more complete understanding of the plasma-enhanced CVD process. In addition, we will discuss reactor configurations and their influence on the films grown. In a later chapter, we will review in some detail the metallic and dielectric films that are being deposited by such techniques.

2.2 PLASMAS

A plasma is characterized as a continuum gas, which is partially ionized and which has equal number densities of electrons and ions (charge neutrality) at each point in the field. When a plasma is subjected to an applied electric field, the electrons can achieve higher energies than heavy particles (atoms, molecules, ions) on average. They can then create many more free radicals by electron impact than would be possible thermally.

2.2.1 Elevated Electron Temperatures in Plasmas

The relationship of the applied electric field to the resulting elevated electron temperature can be seen by a simple analysis. Consider a high velocity electron colliding with an initially stationary heavy particle, as shown below.

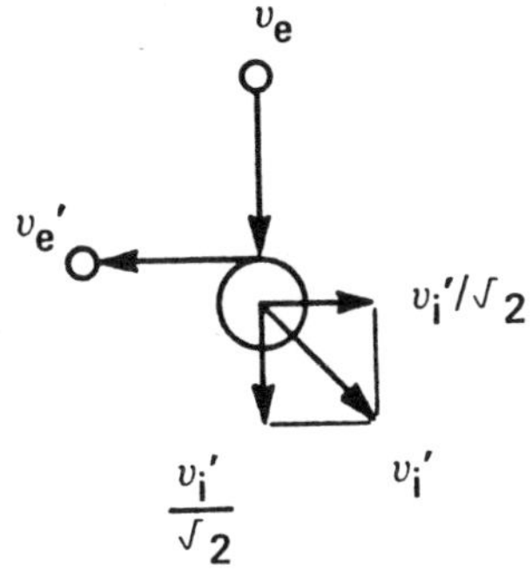

Conservation of energy and axial momentum for this system gives

$$\frac{1}{2} m_e v_e^2 = \frac{1}{2} m_e v_e'^2 + \frac{1}{2} m_i v_i'^2 \quad (1)$$

and

$$m_e v_e = m_i v_i'/\sqrt{2} \quad (2)$$

Replacing v_i' in Equation (1) from Equation (2) gives

$$\frac{1}{2} m_e v_e^2 - \frac{1}{2} m_e v_e'^2 = \frac{1}{2} m_i \left[\frac{m_e^2}{m_i^2} v_e^2 \cdot 2 \right]$$

or

$$\frac{1}{2} m_e v_e^2 - \frac{1}{2} m_e v_e'^2 = \left(\frac{2m_e}{m_i} \right) \frac{1}{2} m_e v_e^2 \quad (3)$$

and the energy an electron loses in an elastic collision with a heavy particle in one collision is $(2m_e/m_i)$ times the electrons' initial energy. Since the electron mass (m_e) is much smaller than the ion mass (m_i), we see that the electron loses very little energy in such collisions. Now, if we think in terms of a continuum with thermal energy, the average particle energy will be $3/2$ kT, where k is the Boltzman constant. Then, the energy lost by an electron in such a collision is

$$\Delta\varepsilon = \frac{2m_e}{m_i} \left[\frac{3}{2} k T_e - \frac{3}{2} k T \right] \quad (4)$$

where T_e is the electron temperature and T is the heavy particle temperature.

The energy gain of an electron in an electric field equals the force on it times the distance it moves. In a time τ, a particle acted on by a force F will move x distance given by

$$x = \left(\frac{F}{m} \right) \frac{\tau^2}{2} .$$

The force on an electron in an electric field is

$$F = eE .$$

So the distance it travels is

$$x = \left(\frac{eE}{m_e} \right) \frac{\tau^2}{2} .$$

Therefore, the energy the electron gains between collisions is

(5) $$\Delta\varepsilon = \frac{1}{2}\frac{(eE)^2}{m_e}\tau^2$$

and we equate Equations (4) and (5) to obtain

(6) $$\frac{1}{2}\frac{(eE)^2}{m_e}\tau^2 = \frac{2m_e}{m_i} \cdot \frac{3}{2}k\,[T_e - T] .$$

For large values of E or τ, we have $T_e \gg T$. Large values of τ correspond to lower pressure regimes, such as are found in glow discharges. The common example is the fluorescent lamp discharge which operates with a low pressure glow discharge in mercury vapor.

2.2.2 Characteristic Parameters in Plasmas

In looking at plasmas in the most general way, we can categorize phenomena in terms of characteristic parameters. For example, a simple gas can be dealt with in terms of a mean-free path, mean collision frequency, or mean thermal energy. If the mean-free path is less than the dimensions of the vessel holding the gas, it can be treated as a continuum. Otherwise, we would have to look at free molecular flow. In contrast, a plasma has many more characteristic parameters than a simple gas.

It will be instructive to consider an electron density (n_e) versus temperature (T) plot[9] for a deuterium plasma, as shown in Figure 1. This plot was developed for a deuterium plasma where the characteristic length is 1 cm. On this plot are displayed lines of constant magnetic field intensity (1, 10^4 and 10^8 gauss). Since we are dealing with a continuum plasma, these are the values of magnetic field necessary for the Lorentz force ($\bar{J} \times \bar{B}$) to equal or exceed the pressure gradient. The main features of the plot are shown by the five regions (S, T, M, EM, and E). They define the electron densities and temperatures where the plasma has scalar or tensor properties and where electric and/or magnetic field effects predominate.

The different regions are delineated by the Debye length (d), the mean-free path (λ), and the Larmor radius r_L). In a plasma, there can be many mean-free paths, since there are many different types of particles (different neutral species, electrons and ions). Of primary interest are the mean-free paths for collisions between electrons and heavy particles (λ_e) and ions and heavy particles (λ_D).

If a magnetic field exists in a plasma, charged particles will tend to gyrate about magnetic field lines between collisions. The radius of the circular path it takes is called the Larmor radius and is expressed as

$$r_L = \frac{v_T m}{qB}$$

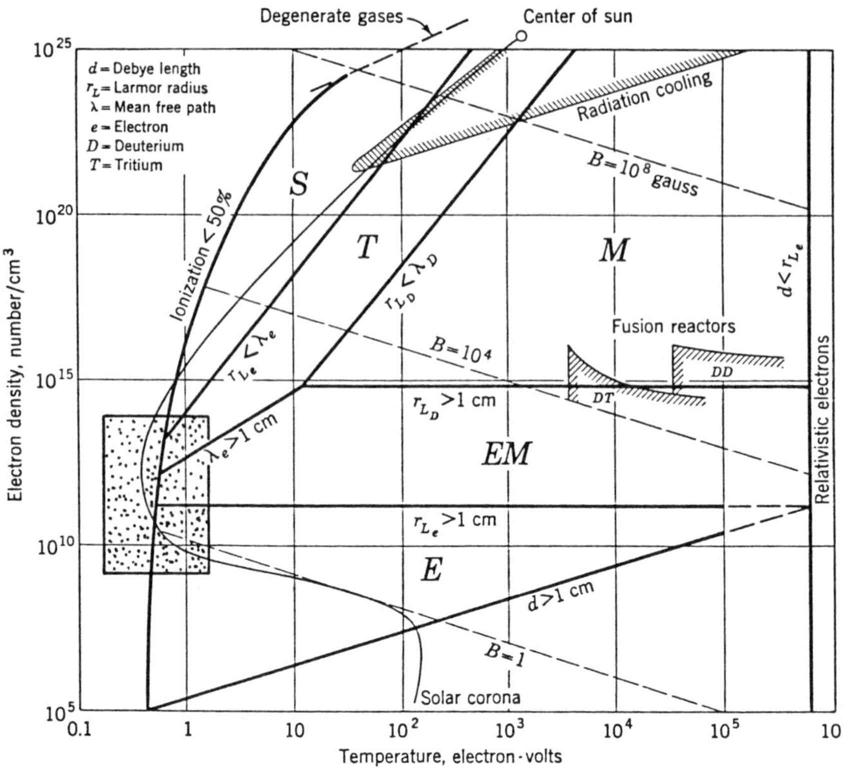

Figure 1: Plasma phenomena displayed for a deuterium plasma.[9]

where v_T is the local velocity of the charged particle normal to the magnetic field line, m is the mass of the charged particle, and q is its charge. Obviously, the mass of an electron is much smaller than the mass of a heavy ion, so the Larmor radius for electrons will be much smaller than for ions. Therefore, if we operate a glow discharge in the presence of a magnetic field, in general the electrons will be confined by the magnetic field, while the ions will not be affected. However, it is difficult to create any significant charge separation in a plasma, so confining the electrons has the effect of confining the ions as well. It is important to note, however, that neutral particles (including free radicals) will not be influenced by the magnetic field.

The Debye length (d) in a plasma is an indication of the distance a strong electric field can extend from a surface into a plasma. It is given by

$$d = \left(\frac{kT\ \varepsilon}{e^2 n_e}\right)^{1/2},$$

where ϵ is the permitivity of free space. For low pressure discharges where n_e can be quite small, the Debye length can be large and a sheath region along a surface (where significant charge separation occurs) can extend a considerable way into the plasma.

The different regions mentioned earlier (S, T, M, EM and E) are defined in terms of these three parameters (λ, d and r_L). Table 1 describes each region and its boundaries.

The region of our n_e versus T plot in which plasma-assisted CVD reactors function is also shown by the shaded area. Clearly, this only represents a very small region of the total space within which a reactor could operate. In the future, operation of CVD reactors in other regions of this plot may lead to CVD films of unique properties.

Table 1: Regions in Deuterium Plasma

Region S	$r_{Le} > \lambda_e$	Collisions occur before an electron can gyrate appreciably; all transport properties are scalar
Region T	$r_{Le} < \lambda_e$, $r_{Ld} > \lambda_d$	A magnetic field will give rise to tensorial properties for electron-related phenomena
Region M	$r_{Le} < \lambda_e$, $r_{Ld} < \lambda_d$	All properties are tensorial
Region EM	$r_{Ld} > 1$ cm	The B field controls the electrons but charge separation and the resulting E field combine with the B field to control the behavior
Region E	$r_{Ld} > 1$ cm, $r_{Le} > 1$ cm	The B field is too weak to influence the plasma motion

It is also of interest to consider a typical plasma used for CVD. If we have p ~ 250 mTorr, we will likely have $T_e \sim 20{,}000°K$. Then, the electron mean-free path and the electron-heavy particle collision frequency can be estimated: and we recognize that the collision frequency is much higher than the highest frequency typically used in a plasma CVD reactor (13.56 MHz). Therefore, electrons will experience many collision during each applied field cycle.

$$\lambda_{ea} \approx \frac{1}{n_a Q_{ea}} = \frac{1}{10^{16}\ \text{cm}^{-3} \times 10^{-15}\ \text{cm}^2} = 1\ \text{mm}$$

and

$$\nu_{ea} \approx \frac{\langle C_e \rangle}{\lambda_{ea}} = \frac{(8kT_e/\pi m_e)^{1/2}}{\lambda_{ea}} = \frac{10^6\ \text{m s}^{-1}}{10^{-3}\ \text{m}} = 10^9\ \text{s}^{-1}$$

2.2.3 Electron Cyclotron Resonance in Plasmas

Among the many phenomena that can occur in a plasma,[11] one of the more interesting from the point of view of discharges used for PECVD is that of electron cyclotron resonance. When a plasma is subjected to an alternating electric field in the presence of a perpendicular static magnetic field, the electrons will receive energy from the electric field but will gyrate because of the magnetic field. Consider the arrangement shown below,

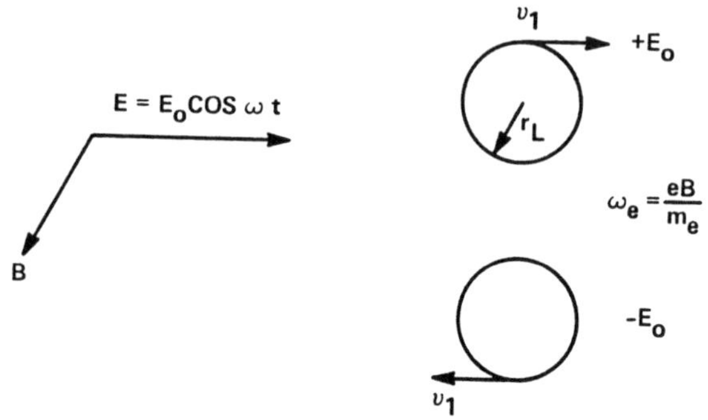

where the magnetic field is normal to the page. Initially, an electron is accelerated to the right by the E field. If $\omega_e = \omega$, however, the B field will turn the electron around just in time to again be accelerated by the E field in the opposite direction. Thus, the electron gains energy as E oscillates in both directions (provided there are many oscillations between collisions), and a resonant condition is achieved. Such a resonance will reduce the electric field necessary to initiate a discharge in a gas,[11] as is shown in Figure 2.

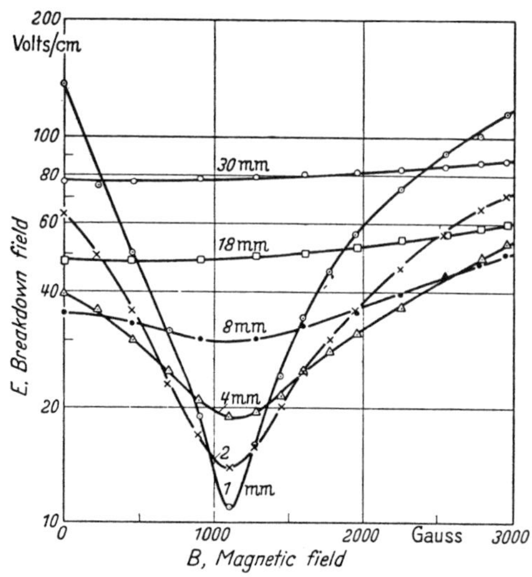

Figure 2: Breakdown field for He + Hg gas as a function of magnetic field for different pressures.[11]

For higher pressures, very little resonance is seen as electron collisions occur so frequently that the electron cannot be turned by the magnetic field in time to catch the reversing electric field. At lower pressures, there is a strong resonance.

The behavior of such a resonant discharge can be described by solving the continuum momentum equations for electron velocity, assuming a constant frequency ν_e. The force on the electrons is both due to the electric field and the Lorentz force caused by the magnetic field. The average power input per unit volume to the plasma is found to be[10]

$$P = \frac{1}{2}\sigma_o E_o^2 \left[\frac{1}{1 + \frac{(\omega - \omega_e)^2}{\nu_e^2}} \right] \tag{7}$$

It is useful to compare this power input to a plasma to the power that would be input when no magnetic field is present ($\omega_e = 0$). Then we can write

$$\tilde{P} = \frac{P}{P_{\omega_e = 0}} \quad \frac{1 + \left(\frac{\omega}{\nu_e}\right)^2}{1 + \left(\frac{\omega}{\nu_e} - \frac{\omega_e}{\nu_e}\right)^2} \tag{8}$$

This relation is plotted in Figure 3 for different values of ω/ν_e. As can be seen, when the frequency of the applied field is large compared to the collision frequency in the plasma, a strong resonance is predicted, in agreement with experimental results.

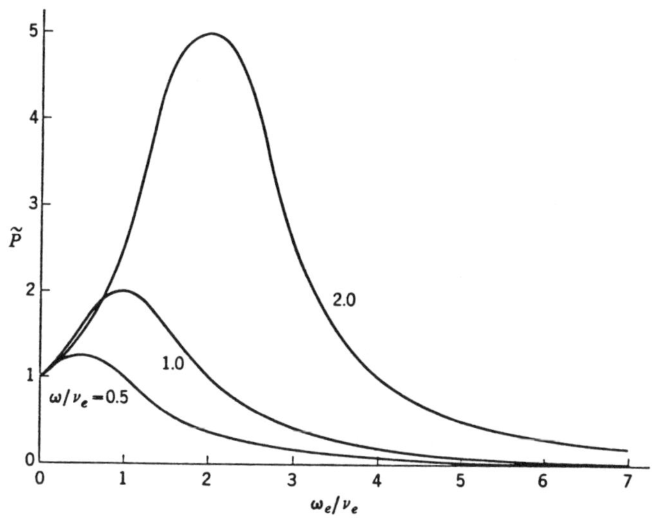

Figure 3: Power input to plasma in the presence of steady magnetic field.[10]

2.3 REACTOR INFLUENCE ON PLASMA BEHAVIOR

In astrophysical studies, one can study plasmas unaffected by solid surfaces. By way of contrast, laboratory plasmas always interact with such surfaces. Accordingly, if we are to properly understand the behavior of laboratory plasmas, we must inquire into the nature of the plasma-solid surface interaction.

There are several aspects of this interaction that we will touch on. First, we will review concepts of the DC and AC discharge. Then, the consequences of using unequal size electrodes will be discussed. For AC discharges, frequency will also play a role; and finally, the influence of magnetic fields on discharges will be considered.

2.3.1 DC/AC Glow Discharges

A glow discharge in a low pressure gas (~1 Torr) created by a DC applied voltage exhibits a nonuniform appearance. A typical discharge is shown in Figure 4.[12] Since the cathode is cold, the discharge is maintained by secondary electrons produced there by positive ion impacts. The ions experience a strong electric field near the cathode, which causes them to accelerate toward it. The sheath is the region next to the cathode in which charge neutrality is not obeyed and relatively few collisions occur. This encompasses the Aston, Crookes and Faraday dark spaces, and the cathode and negative glow regions. There is an excess of ions in this region, hence the net positive charge there. The positive column has no net space charge. Therefore, it is the plasma we referred to earlier. It is of a high electrical conductivity, so a relatively modest electric field is all that is necessary to conduct the DC current through it. Ions and electrons in this region can be lost by gas phase recombination, or diffusion to the tube walls. They can be regenerated by electron impact ionization in the positive column, or the secondary electron emission from the cathode mentioned earlier. A similar but much smaller sheath appears at the anode.

There is also a potential difference between the positive column and tube wall. This potential difference is created because the electrons are much more mobile than heavy ions and tend to flow rapidly out toward any bounding surface. Since the tube wall is an insulator, they tend to collect there causing the insulator to assume a negative potential relative to the plasma. This creates an electric field close to the tube wall which hinders further electron flow towards it. A deficit of electrons forms in a sheath close to the surface, and this sheath assumes a net positive charge. Ions in the plasma, however, see the tube wall potential which is negative compared to the plasma and are attracted to it. This is the diffusion to the tube walls mentioned in the previous paragraph, and is often referred to as "ambipolar" diffusion.

If the glow discharge of Figure 4 is operated under alternating voltage conditions, we observe a discharge with two dark spaces. This is in reality a series of DC discharges of alternating polarity. Up to about 10 kHz, the frequency is low enough so that the discharge lights and extinguishes on each cycle. There is sufficient time between cycles for most electrons to leave the positive column and be lost to the tube walls. The loss of electrons extinguishes the glow discharge. Above 10 kHz, there is not enough time for the electrons in the positive column to be lost to the walls, so the discharge remains lit continuously. Depending on the discharge geometry and gas involved, the starting voltage for an AC discharge can depend on the RF frequency and pressure.[11]

Finally, when an AC discharge is operated with a blocking capacitor between the power supply and one of the electrodes, that electrode will assume negative self bias. Such an average negative voltage on this electrode can serve to accelerate ions toward it with considerable energies.

It is interesting to note that this is the ion bombardment that is used in plasma ("dry") etching to promote the anisotropic character of that process. It is also important in the understanding of sputtering phenomena.[13]

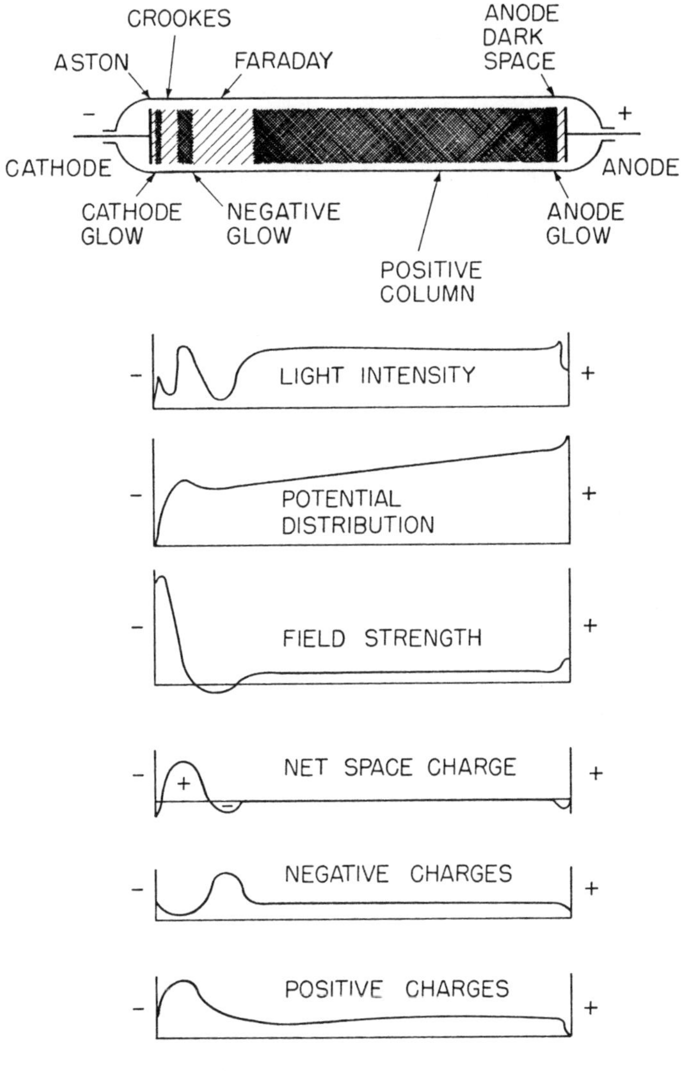

Figure 4: A DC glow discharge at low pressure.[12]

Why the negative self bias forms has been described by Butler and Kino[14] and can be seen from Figure 5. In the first figure, an alternating potential is applied to a conducting probe in a plasma. When the probe sees a positive voltage, a large electron current flows. Reversing the voltage produces only a small current flow due to the immobility of heavy ions. Thus when the probe is conducting, the self bias is zero and large net currents flow on average. On the other hand, when the probe is nonconducting (i.e., electrode attached to the blocking capacitor), the behavior is shown in the second figure. In order for the average current to be zero, the average applied voltage (i.e., RF signal) must become negative.

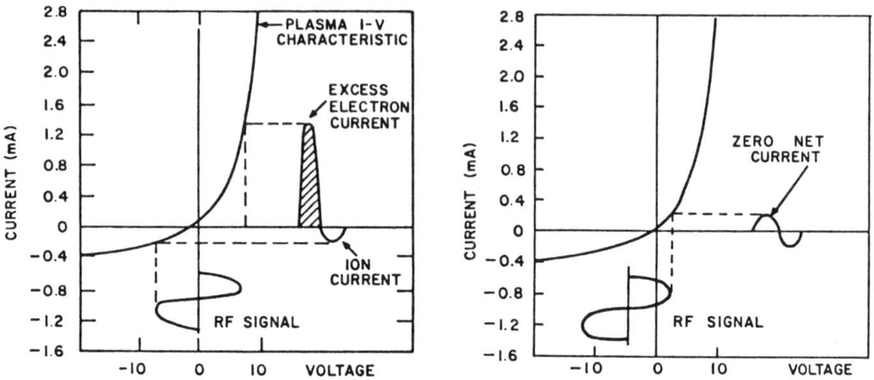

Figure 5: Creation of negative self bias in AC discharge.[14]

2.3.2 AC Discharges with Unequal Area Electrodes

Next, it will be valuable to consider the discharge behavior when the electrodes are not of equal size. A simplified analysis of this situation can be made[15] if we make a number of approximations. Consider a geometry such as shown in Figure 6, where a blocking capacitor is used between the power supply and electrode 1. The function of the blocking capacitor is to allow a DC bias to exist between the DC plasma potential and the electrode adjacent to the capacitor.

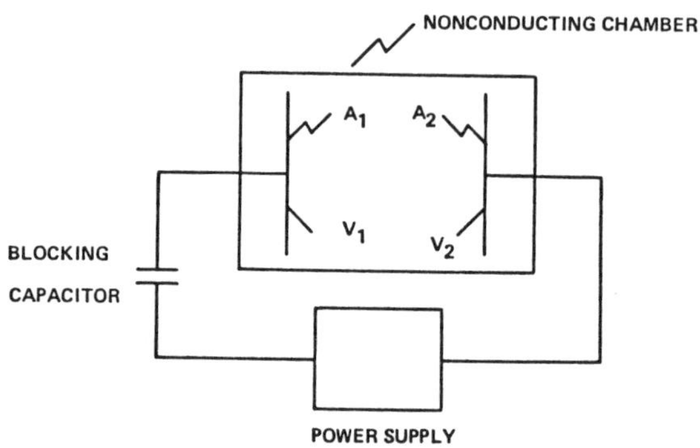

Figure 6: Reactor with unequal size electrodes.

We begin by assuming that the ion current density to all internal surfaces is equal. Then, if we assume a collision-free sheath with the field being zero at the edge of the positive column, we can express the space charge-limited ion current as[16]

$$j_i \propto \frac{V^{3/2}}{L^2 m_i^{1/2}} \quad ,$$

where j_i is the ion current, V is the potential difference across the sheath, L is the sheath thickness, and m_i is the ion mass.

Since the ion currents on each electrode have been taken equal, we can write

$$\text{(9)} \qquad \frac{L_1}{L_2} = \left(\frac{V_1}{V_2}\right)^{3/4}$$

For capacitances in series, we can write

$$\text{(10)} \qquad V_1C_1 = V_2C_2 \ ,$$

where C_1 and C_2 are the capacitances of the two sheaths. Again, we can assume the capacitances of a dark space to be $\alpha A/L$, where A is the electrode area. Then, we have

$$\text{(11)} \qquad \frac{C_2}{C_1} = \frac{A_2L_1}{A_1L_2} \ .$$

Eliminating L_1/L_2 between Equations (9) and (11) gives

$$\text{(12)} \qquad \left(\frac{V_1}{V_2}\right)^{3/4} = \frac{A_1C_2}{A_2C_1} \ .$$

Then, replacing C_2/C_1 in Equation (12) from Equation (10) results in the desired relationship between V_1/V_2 and A_1/A_2.

$$\frac{V_1}{V_2} = \left(\frac{A_1}{A_2}\right)^4 \ .$$

Clearly, there are many approximations in this model, so close agreement with experimental data should not be expected. It is interesting, however, to consider some relevant experimental data.[17] In these experiments, peak-to-peak and DC bias voltages were measured for a 13.56-MH_Z RF glow discharge confined within a reactor chamber with two electrodes of unequal area. Using a simplified theoretical model of the discharge, the authors derived values for the voltage on each electrode relative to the plasma potential, and could then calculate the voltage ratios from the measured data. It was found that the voltage ratios depended not only on the reactor area ratio, but also on the electrode material, the peak-to-peak voltage across the electrodes, the gas and its pressure. Results are shown in Figure 7 for Argon gas in a stainless steel system at 50 mTorr with a 600 peak-to-peak voltage.

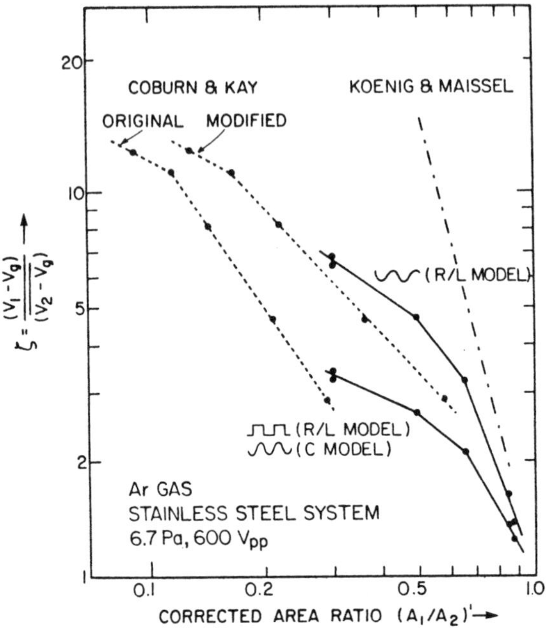

Figure 7: Voltage ratio versus area ratio for argon plasma.[17]

For these particular conditions, it can be seen that $V_1/V_2 = (A_1/A_2)^n$ is a reasonable model, but that $n \cong 4$ only between $A_1/A_2 = 0.6$ and 1.0. For smaller area ratios, we find $n \cong 1$. The data reported by Coburn and Kay[18] were modified by the author by correcting the areas they used in their calculation.

From these results, we see that under certain conditions it is possible to have substantial ion bombardment on the powered electrode while almost none on the remainder of the chamber. The magnitude of the DC bias on each electrode will, in general, depend on peak-to-peak voltage, gas chosen, pressure and chamber materials, as noted earlier. For a given experimental setup, we can change the DC bias on the grounded electrode of Figure 6 by placing a variable LC circuit in between it and ground.[19] Such an arrangement where the second electrode is isolated from the grounded outer chamber is shown in Figure 8. With this circuit, it was possible to develop a 30V DC bias on the substrate holder.

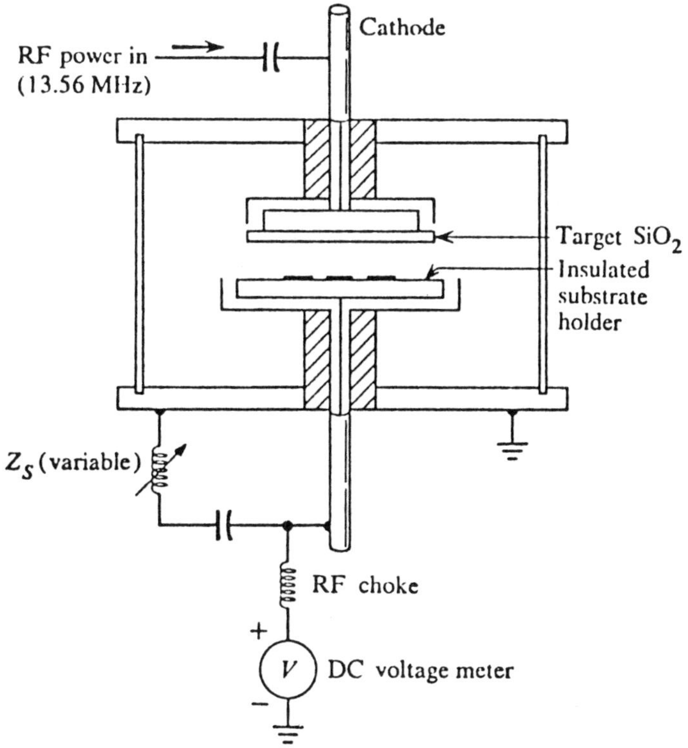

Figure 8: System for substrate electrode tuning to decrease DC bias.[19]

2.3.3 Frequency Effects on RF Plasma Reactor Behavior

When the RF discharges are used to create plasmas in PECVD reactors, the influence of the frequency at which the discharge is operated is another question that must be explored. It is generally recognized that a lower frequency discharge (50 to 100 kHz) will produce a CVD film with greater compressive stress than a film created in a higher frequency discharge (13.56 MHz). The speculation has been that the ion bombardment is more intense at lower frequencies, and this bombardment causes the film stress to be compressive. In fact, recent careful experimental work has verified the more intense ion bombardment at low frequencies.[20] Studies were carried out in a plasma etching apparatus where a hole in one electrode allowed ions to be examined in a mass spectrometer while different retarding voltages were applied. The retarding voltage necessary to cut off beam current was recorded as a function of both power level and frequency. The results are shown in Figure 9, where earlier results[21] are shown in parenthesis.

There are two potential explanations of why the ion bombardment is more intense at low frequencies. First, the sheath potential drop, on average, will be higher at the lower frequencies. The electrons are lighter than ions, so they

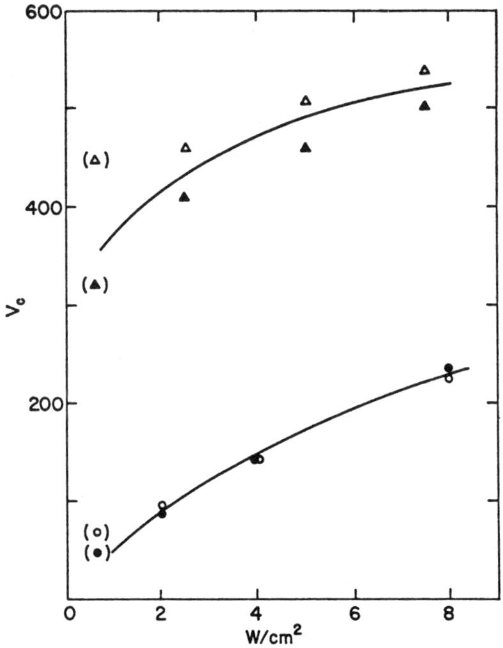

Figure 9: Cl_2 plasma beam maximum ion energy. Circles = 27 MHz; triangles = 100 kHz; solid = Cl_2^+; open = Cl^+.[20] Reprinted by permission of the publisher, The Electrochemical Society, Inc.

tend to preferentially diffuse out of the plasma, and the electrode assumes a negative bias. As frequency is increased, there is increasingly less time available for charged particles to diffuse to the reactor walls between cycles. Therefore, if there is less opportunity for electrons to diffuse out at higher frequencies, there will be less need for a strong bias to form. If we recognize that the negative bias formed has the effect of accelerating ions toward the surface, then more bias (lower frequencies) means more ion bombardment on the forming film.

Secondly, the plasma potential will vary with time, depending on the cycle of the applied potential.[22] If the ion can transit the sheath before the applied electric field reverses, it can experience the maximum sheath potential. As the frequency is raised, the ion cannot cross the sheath before the field reverses, so it experiences the average sheath potential which is approximately one-third the maximum. Therefore, ion bombardment will be less intense at the highest frequencies.

2.3.4 Influence of Applied Magnetic Fields on RF Plasma Reactors

The final aspect of reactor influence on a plasma that we wish to discuss is the use of an applied magnetic field. This is one more parameter that can be adjusted to modify the plasma and the reactive species produced. For the mean-free paths typical of plasmas used in PECVD equipment (~1 mm), a magnetic field on the order of 100 gauss is sufficient to cause the electrons to gyrate many times between collisions. This has the effect of immobilizing the electrons. Since there can be very little charge separation in a plasma, the ions will be attracted by the immobilized electrons and be immobilized as well. All neutrals (free radicals, excited species, etc.) will be uneffected by the magnetic field. Therefore, the magnetic field provides a means for possibly confining the plasma to one region of the reactor (over the wafer, for example) or modifying the spatial distribution of reactive species.

Such effects were demonstrated in an experiment where a discharge in CF_4 gas at 10 mTorr pressure was established along the axis of a cylindrical tube, and an axial magnetic field of 30 gauss was applied.[23] The plasma was confined to the central 1-inch diameter core within the 5-inch diameter tube. Two new phenomena are created by this arrangemement. First, the species that cause etching (i,e., of Si or SiO_2) are neutrals (in this case, F atoms created by $CF_4 + e^- \rightarrow CF_3^+ + F + 2e^-$), and they are present throughout the region. If ions enhance the etch rate, they are only present in the plasma, so the etch rate on a wafer will depend on whether it is in the plasma or not. Again, when the wafer is in the plasma, biasing it to increase ion bombardment will also increase etch rates. Second, the chemical composition of the gas will differ because the plasma region will be carbon rich due to the retention of CF_3^+ species, while fluorine atoms are free to drift away. The excess carbon can lead to deposition rather than etching.

Results shown in Figure 10 illustrate this behavior where deposition is seen at zero bias in the plasma, but etching is observed when −40V bias is applied.

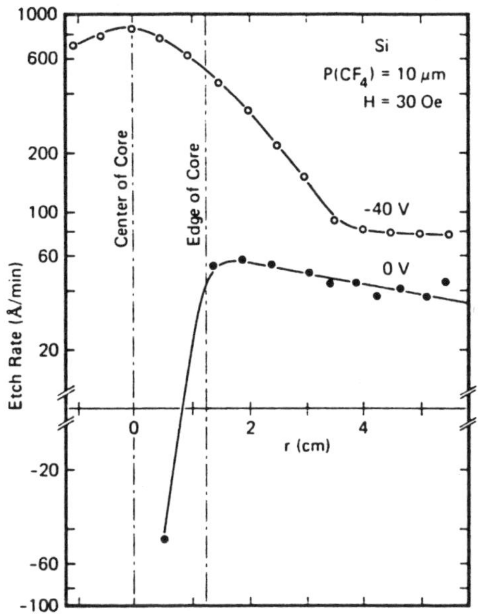

Figure 10: Etch and deposition behavior in a magnetically-confined cylindrical plasma.[23]

2.4 PLASMA-ENHANCED CVD (PECVD) REACTORS

Having covered some of the elements of plasma behavior and how it relates to reactors, it is appropriate to consider the plasma-enhanced CVD reactor specifically. This is where the plasma is created in an appropriate gas mixture so that a suitable thin film will grow on a chosen substrate. As discussed in the previous chapter, a gas mixture can react thermally, both in the gas phase and on the surface, to grow films. A similar process occurs with plasma enhancement, except that the gas mixture presented to the surface has many more species due to decomposition of the starting gas by high-energy electron impact, and there can be a high density of such species.

There are basically three ways to create such a plasma for purposes of thin film deposition. These are shown in Figure 11. In the first, a pair of conducting electrodes are exposed to the low-pressure reacting gas and a DC or AC glow discharge is created. If a metallic film is being deposited, either a DC or AC discharge can be used. If a dielectric is being deposited, one must use an AC discharge, because the metal electrodes will become coated and a DC discharge would extinguish. The second approach uses a coil wound around a tube containing the reacting gas. When an AC current flows through the coil, an alternating electric field is induced within the tube and causes the gas to break down. Finally, if a pair of conducting electrodes are situated outside the tube, as in Figure 11B, and an AC potential is applied to them, the electric field is felt within the tube and again a discharge is created.

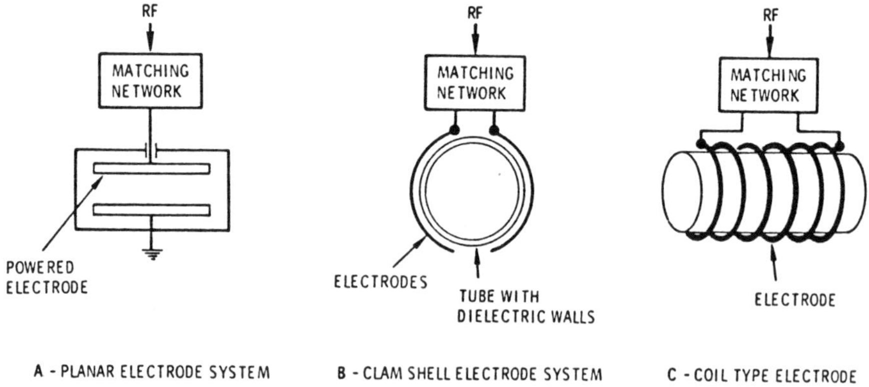

Figure 11: Geometries of plasma-assisted CVD reactors: (A) parallel-plate discharge, (B) tube with capacitive coupling, (C) tube with inductive coupling.[13]

Tube reactors are generally used for resist ashing or less critical depositions. In resist ashing, wafers are inserted into an oxygen plasma which reacts with (ashes) the hydrocarbon-based resist to form gaseous products (CO, CO_2, etc.). These reactors are simple and relatively inexpensive to build. It is difficult to have them etch uniformly on many wafers, but this is not a critical issue for resist ashing.

2.4.1 Cold-Wall, Parallel-Plate PECVD Reactors

The original plasma-enhanced CVD reactor was developed by Reinberg[24] and is illustrated in Figure 12. This was a parallel-plate reactor of circular symmetry, where the wafers sat on a heated platen. The reactants were introduced at the outer periphery, and the exhaust was at the center. Reinberg theorized that the discharge intensity would be higher in the center, tending toward higher deposition rates there. Offsetting this would be the higher flow velocities in the center (shorter residence times), leading to uniform deposition rates from center to outer edge. Based on this concept, a patent for this reactor was issued.[25]

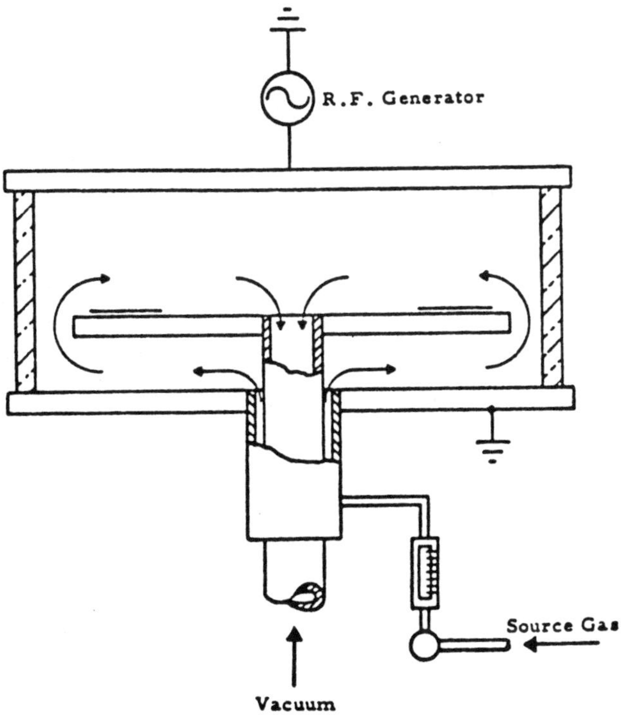

Figure 12: Radial-flow, plasma-enhanced CVD reactor after Reinberg.[24]

In an attempt to develope their own concept, Applied Materials built a reactor which introduced the reactants in the center and exhausted at the periphery.[2] This design is shown in Figure 13.

The Applied Materials reactor is fabricated of aluminum (including the wafer-holding susceptor), and the susceptor is rotated by a magnetic coupling. Because of this rotation, the susceptor must be heated by radiation, typically to 325°C. Reactant gases enter at the center and flow outward where they are exhausted. Since the electrodes are 2 inches apart and approximately 26 inches in diameter, there is a relatively uniform glow discharge between them. In spite of Reinberg's predictions, deposition is quite uniform with radius.

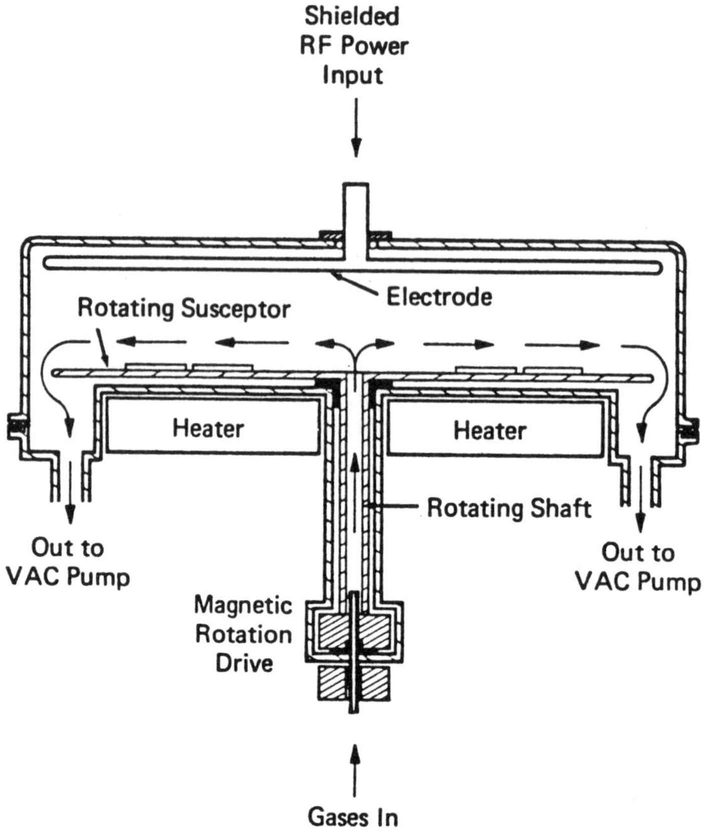

Figure 13: Radial-flow, plasma-enhanced CVD reactor.[2]

The platen is grounded, and the upper electrode is powered with a low-frequency rf power supply (~50 kHz) at a power level of 500 to 1000 watts. The reactor operates in the batch mode with a wafer load of twenty-two 4-inch wafers, for example. For larger wafers, the load size is less. Although the load size is restricted, high quality films are produced.[7]

A recent modification of the Applied Materials system incorporates a perforated upper electrode for more uniform gas distribution, as shown in Figure 14. Such a design change was necessary when deposition with 2% SiH_4 in N_2 was attempted. Because of the low concentration of SiH_4, the reactant became depleted as gases flowed outward and deposition became quite non-uniform. This problem was corrected when the reactant gases were introduced more uniformly with radius. The motivation behind the use of 2% SiH_4 in N_2 was safety. It is felt that such a dilute mixture cannot sustain an explosive reaction in a gas cabinet.

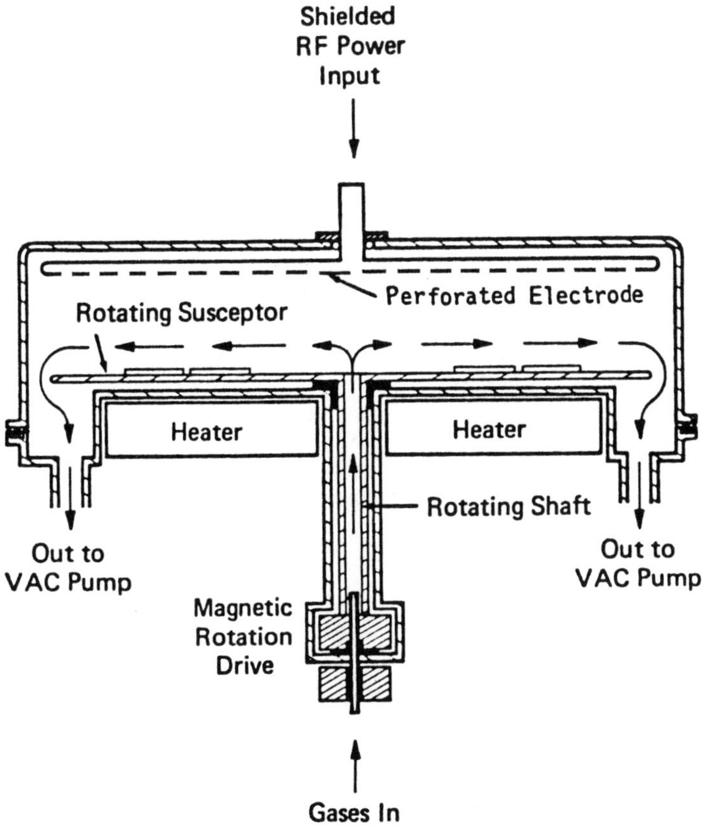

Figure 14: Radial flow plasma-enhanced CVD reactor with perforated upper electrode for uniform reactant gas introductions.

2.4.2 Hot-Wall, Parallel-Plate PECVD Reactors

The two reactors just described are parallel plate reactors. However, they are also cold wall reactors. In other words, the electrode holding the wafers is hot, but all other surfaces exposed to the plasma are cold, or at least not heated. This is done to minimize the deposition on other surfaces so that down time for cleaning can be kept as short as possible.

The same reactor concept would be valid in a hot wall system, if the entire reactor were placed in a furnace. In this way, temperature uniformity would be excellent by definition. Obviously, this would be awkard and not economically attractive. However, if our electrode geometry were of two parallel, narrow rectangular electrodes, the structure could conveniently fit into a hot tube (much like a diffusion furnace). In this way, a large batch load could be run in a relatively hot furnace tube. Such a hot wall system is shown in Figure 15.

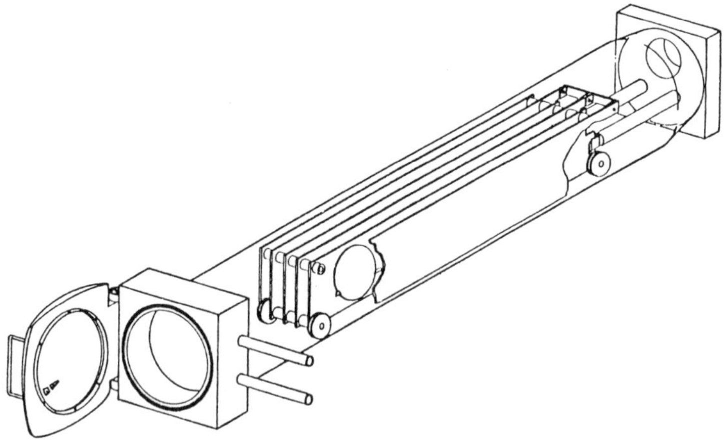

Figure 15: Hot-wall, parallel-plate reactor for plasma-enhanced CVD. (Courtesy of Pacific Western Systems, Inc.)

Multiple reactangular electrodes are arranged so that they fit down the length of a tube and are alternately powered by a 400-kHz power supply. The electrodes are fabricated of graphite. A major attraction of the hot wall system is the large wafer load that can be run (i.e., 84 4-inch wafers) at one time. This is offset to some extent by the fact that the electrode structure cools off each time wafers are unloaded, and the time needed to reheat upon insertion into the furnace detracts from wafer throughput.

2.5 NOVEL PLASMA-ENHANCED CVD REACTORS

Current commercial plasma-enhanced CVD reactors operate with only two physical concepts. In one case, we have the inductively-excited discharge in a tube, which is used for plasma ashing of resist. The other is the parallel plate arrangement using high-frequency RF power to create a low-pressure glow discharge, where the wafers to be coated sit on one of the electrodes.

In reality, these are only two of many arrangements that could be devised to create and deliver to a substrate large quantities of reactive species using a plasma. Since there are many shortcomings to existing commercial plasma-enhanced CVD reactors, it will be useful to explore other reactor concepts that are under development, but have yet to be widely developed commercially. Whether or not they will lead to practical production systems remains to be seen.

2.5.1 Electron Cyclotron Resonance (ECR) CVD Reactor

One approach being pursued by Japanese investigators makes use of the electron cyclotron resonance phenomena discussed earlier.[24] In this case, a

2.45-GHz microwave generator feeds microwave energy into a rectangular wave guide and then through a quartz window into a plenum chamber, as shown in Figure 16.

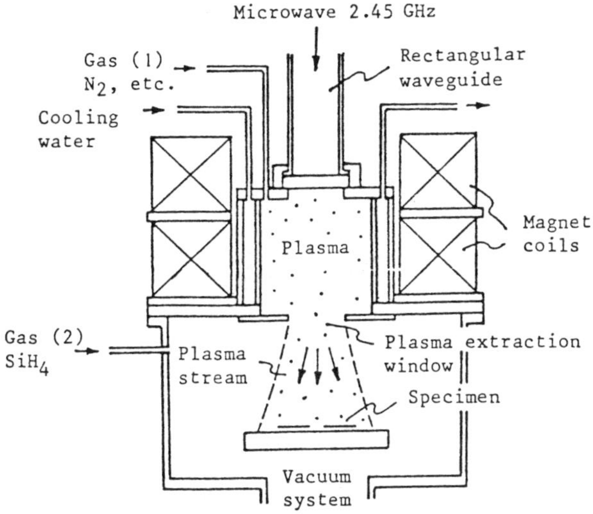

Figure 16: ECR (Electron Cyclotron Discharge) reactor for plasma-enhanced CVD (after Matsuo[26]).

A very low pressure gas is introduced into this plenum and a magnetic field is established by a solenoidal magnetic coil placed outside this chamber. The very high frequency electric field established by the microwave source ionizes the reactant gas to a small extent. However, when the steady magnetic field is applied, a resonance condition is achieved (electron cyclotron resonance) and the energy transfer from the microwave source to the plasma is maximized. Such strong resonant interaction causes a much higher degree of ionization, dissociation, and excitation than would be possible with the microwave energy alone.

It is important to recognize that in this system the intense degree of ionization and dissociation is established in a region away from the wafer. These gases are then fluid dynamically transported to the wafer where deposition occurs. If the wafer had been present in the region where the plasma was being generated, serious damage would have been done to it. It is the separation of the generation and deposition processes that makes this concept feasible.

For the deposition of silicon nitride, the reactant gases were SiH_4 and N_2 flowing at a pressure of 0.1 mTorr. With the excitation frequency fixed at 2.45 GHz, a steady magnetic field of 875 gauss produced the condition of electron cyclotron resonance. The plasma produced by this condition then flows along a divergent magnetic nozzle until it impinges on the wafer being coated. The wafer which is electrically floated in the chamber charges negatively (~15V) and causes some ion bombardment of the growing film.

Silicon nitride films produced by this technique proved to be comparable to those created in parallel-plate reactors in terms of stoichiometry and hydrogen content, as evidenced by the data presented in Figures 17 and 18.

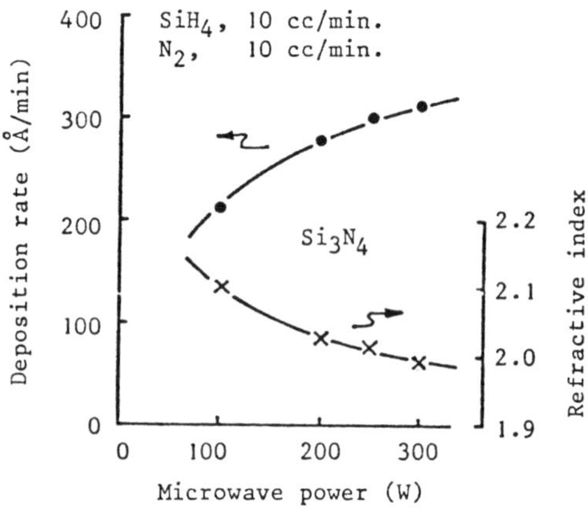

Figure 17: Deposition rate and refractive index for silicon nitride films as a function of microwave power for ECR reactor (after Matsuo[26]).

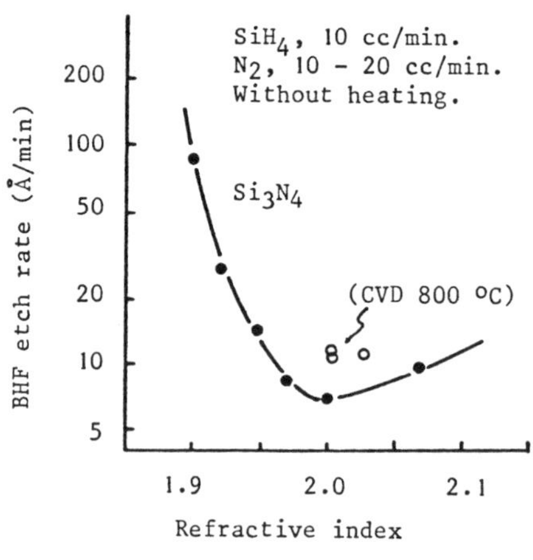

Figure 18: BOE etch rate versus refractive index for silicon nitride films for ECR reactor (after Matsuo[26]).

Silicon nitride films deposited in parallel-plate, plasma-enhanced CVD reactors will be discussed in greater detail in a later chapter. However, they typically have a refractive index on the order of 2.0, partly because of hydrogen incorporated into the layer, and the ECR films appear similar. Also, as will be discussed again later, the buffered oxide etch (BOE) rate for such films is some indication of their hydrogen content, high rates indicating very high atomic percentages of hydrogen. Etch rates for silicon nitride films produced in parallel-plate reactors will frequently range up to 100 Å/min, so the ECR films appear to be of good bulk quality.

There are a number of interesting observations that can be made regarding these results. First, the deposition rate achieved in these experiments is comparable to those obtainable by conventional PECVD, in spite of the fact that the ECR unit operates at 0.1 mTorr and the conventional unit at 300 mTorr. This indicates a much higher degree of excitation of the ECR plasma.

In spite of this, we are still comparing a single-wafer CVD machine to the batch parallel-plate units, both with the same deposition rates. Therefore, it is unlikely such a machine will be of much interest for production of films for integrated circuit passivation.

In addition, due to the low plasma pressure and guiding magnetic field, such a reactor deposits in an anisotropic fashion. In other words, the film will either be thicker or denser on a horizontal surface exposed to the ECR plasma than it will be on vertical surfaces. This makes such films of limited value for coating integrated circuits with the usual topography.

2.5.2 Parallel Electrode, Hot-Wall PECVD Reactor

Instead of arranging the electrodes as long, narrow rectangular plates positioned along the length of a hot tube (see Figure 15), it would be highly desirable to arrange them normal to the tube axis, as shown inFigure 19.

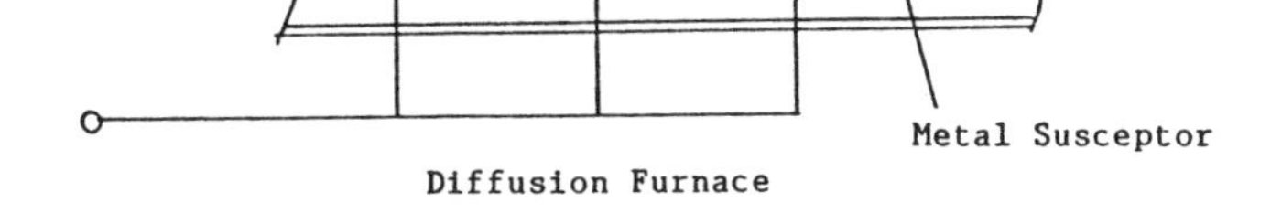

Figure 19: Hot tube parallel-plate PECVD reactor.

The great advantage of this arrangement over the one currently used commercially is that for a given tube size, many more wafers could be loaded in each batch. Unfortunately, although a number of U.S. and Japanese equipment vendors have tried to develop such a design, it has not proven practical. Major difficulties were experienced due to the complexity of the electrode structure and in maintaining uniform discharges along the side of the tube.

2.5.3 Ionic Systems Concept

In order to carry out plasma-enhanced silicon nitride deposition with low hydrogen content and on room-temperature wafers, this small company has developed a system in which nitrogen is ionized in an RF (13.56 MHz) glow discharge in one chamber, and then the excited gas flows into a second chamber where silane is added. The mixture then flows over room-temperature wafers. Since excited nitrogen is metastable, it can remain excited as it flows into the second chamber where it can decompose the silane. This then provides the reacting mixture necessary to grow silicon nitride films at rates of 150 to 300 Å/min. Again, separation of generation and deposition has permitted unusual results to be achieved.

REFERENCES

1. Sherman, A., Plasma-assisted chemical vapor deposition processes and their semiconductor applications. *Thin Solid Films* 113:135 (1984).
2. Rosler, R.S., Benzing, W.C. and Baldo, J., A production reactor for low-temperature plasma-enhanced silicon nitride deposition. *Solid State Technology* 19(6):45 (1976).
3. Carlson, D.E. and Wronski, C.R., Amorphous silicon solar cell. *Appl. Phys. Lett.* 28:671 (1976).
4. Wroge, D.M. and Hess, D.W., Plasma-enhanced deposition of iron/iron oxide films. *Proceedings of the Symposium on Plasma Etching and Deposition.* Electrochemical Society, Pennington, NJ, 81-1:30 (1981).
5. Reif, R., Plasma enhanced chemical vapor deposition of thin crystalline semiconductor and conductor films. *J. Vac. Sci. Technol.* A2(2):429 (1984).
6. Hess, D.W., Plasma-enhanced chemical vapor deposition of metal and metal silicide films. *Mat. Res. Soc. Symp. Proc.,* Vol. 38, 1985, Materials Research Society.
7. Van de Ven, E.P.G.T., Plasma deposition of silicon dioxide and silicon nitride films. *Solid State Technol.* 24(1):167 (1981).
8. Chapman, B., *Glow Discharge Processes,* John Wiley & Sons, NY (1980).
9. Fishman, F.J., Kantrowitz, A.R. and Petschek, H.E., Magnetohydrodynamic shock wave in a collision-free plasma. *Rev. Mod. Phys.* 32:959 (1960).
10. Sutton, G.W. and Sherman, A., *Engineering Magnetohydrodynamics,* McGraw-Hill, NY (1965).
11. Brown, S.C., Breakdown in Gases: Alternating and High-Frequency Fields. In *Handbuch der Physik,* Vol. 22, ed. S. Flugge (1956) Springer-Verlag.
12. Brown, S.C., *Basic Data of Plasma Physics,* John Wiley & Sons, NY (1959).
13. Thornton, J.A., Plasmas in Deposition Processes. In *Deposition Technologies for Films and Coatings,* ed. Bunshah, R.F., Noyes Publications, NJ (1982).
14. Butler, H.S. and Kino, G.S., Plasma sheath formation by radio-frequency fields, *Phys. Fluids* 6:1346 (1963).

15. Koenig, H.R. and Maissel, L.I., Application of RF discharges to sputtering. *IBM J. Res. Develop.* 14:168 (1970).
16. Francis, G., The Glow Discharge at Low Pressure. In *Handbuch der Physik,* Vol. 22, ed. S. Flugge (1956) Springer-Verlag.
17. Horwitz, C.M., RF sputtering-voltage division between two electrodes. *J. Vac. Sci. Technol. A,* 1(1):60 (1983).
18. Coburn, J.W. and Kay, E., Positive-ion bombardment of substrates in RF diode glow discharge sputtering. *J. Appl. Phys.* 43:4965 (1972).
19. Logan, J.S., Control of RF sputtered film properties through substrate tuning. *IBM J. Res. Develop.* 14:172 (1970).
20. Smith, D.L. and Bruce, R.H., Si and Al etching and product detection in a plasma beam under ultrahigh vacuum. *J. Electrochem. Soc.* 129: 2045 (1982).
21. Bruce, R.H., Ion response to plasma excitation frequency. *J. Appl. Phys.* 52:7064 (1981).
22. Bruce, R.H., Frequency dependence of CCl_4 etching. *Proceedings of the Symposium on Plasma Etching and Deposition,* Electrochemical Society, Pennington, NJ, 81-1:243 (1981).
23. Minkiewicz, V.J., Chen, M., Coburn, J.W., Chapman, B.N. and Lee, K. Magnetic field control of reactive plasma etching. *Appl. Phys. Lett.* 35:393 (1979).
24. Reinberg, A.R., RF Plasma deposition of inorganic films for semiconductor applications. Electrochem. Soc. *Extended Abstracts Volume* 1974-1, pg. 21. This figure was originally presented at the Spring 1974 Meeting of The Electrochemical Society, Inc. held in San Francisco, California.
25. Reinberg, A.R., Radial Flow Reactor, U.S. Patent 3,757,733, Sept. 11, 1973.
26. Matsuo, S. and Kiuchi, M., Low temperature deposition apparatus using an electron cyclotron resonance plasma. *Proc. Symp. on Very-Large-Scale Integration Science and Technology,* Electrochemical Society, Pennington, NJ, pg. 83 (1982). These figures were originally presented at the Fall 1982 Meeting of The Electrochemical Society, Inc. held in Detroit, Michigan.

3

Thermal CVD of Dielectrics and Semiconductors

3.1 INTRODUCTION

In the present chapter, we will turn our attention to films deposited by thermal CVD that are either dielectrics or semiconductors. There are, as one would expect, many films that can be deposited by this technique. In addition, there are many gaseous reactants that one can use to create each film, the choice depending on the film characteristics desired. Rather then attempt to catalogue all of the possible films and reactants, we will choose instead to focus on silicon dioxide, silicon nitride, polysilicon, and epitaxial silicon as the films of interest. At the same time, we will only look at those reactant gases that have been used for integrated circuit manufacture. An excellent survey of the film types that can be deposited by CVD and the many reactants that have been used to obtain them has been given by Kern.[1]

3.2 SILICON DIOXIDE

Silicon dioxide films have been an essential factor in the manufacture of integrated circuits from the earliest days of the industry. They have been used as a final passivation film to protect against scratches and to getter mobile ion impurities (when doped with phosphorus). Another application has been as an interlayer dielectric between the gate polysilicon and the aluminum metalization. Initially, most such films were deposited in atmospheric pressure systems. In recent years, low pressure processes have assumed greater importance. We will begin by examining the atmospheric process.

3.2.1 Atmospheric Pressure

Although the atmospheric pressure SiO_2 film deposition process was the first CVD process used, it continues in use today because it can be run successfully at low temperatures (400° to 450°C). Therefore, it can be used as a final passivation film over aluminum, which would be damaged with processing temperatures over 500°C.

If SiH_4 is mixed with O_2, using an oxygen to SiH_4 ratio above 3:1, and this mixture is heavily diluted with an inert gas, then SiO_2 will be deposited on a hot plate at temperatures above 240°C.[2] The typical reactor is a cold-wall type where the wafer holder is heated. The walls are cooled to try to minimize the deposition on them so that reactor cleaning is kept to a minimum. Several commercial reactors are available that implement this process, and they will be reviewed in Chapter 6.

If the SiH_4/O_2 mixture is not sufficiently diluted with an inert gas, then gas phase nucleation typically occurs and SiO_2 particulates are formed. Generally, N_2 is used as the diluent, but some work has been done with Ar, CO_2, and He. Depending on the reactor configuration, an inert gas effects the deposition rates in different ways.

For most applications of CVD SiO_2, the oxide is doped with PH_3 so that a film of phosphosilicate glass is formed. This film is not a mixture of P_2O_5 and SiO_2, but rather a chemical compound of composition $(SiO_2)_x \cdot (P_2O_5)_{1-x}$. It is also possible to deposit borosilicate glass using B_2H_6 rather than PH_3. Finally, an interesting new material fabricated using both PH_3 and B_2H_6, borophosphosilicate glass, will be considered later. For the moment, we will restrict our attention to phosphosilicate glass films.

In order to see more clearly the effect of deposition temperature and oxygen/hydride ratio (hydride = SiH_4 + PH_3) on film deposition rate, we consider the three-dimensional plot shown in Figure 1. As expected, increasing the deposition temperature increases the film deposition rate. On the other hand, too high an O_2-to-hydride ratio actually decreases the deposition rate. It is speculated that the excess O_2 adsorbs onto the surface and prevents the SiH_4 from adsorbing and decomposing there.

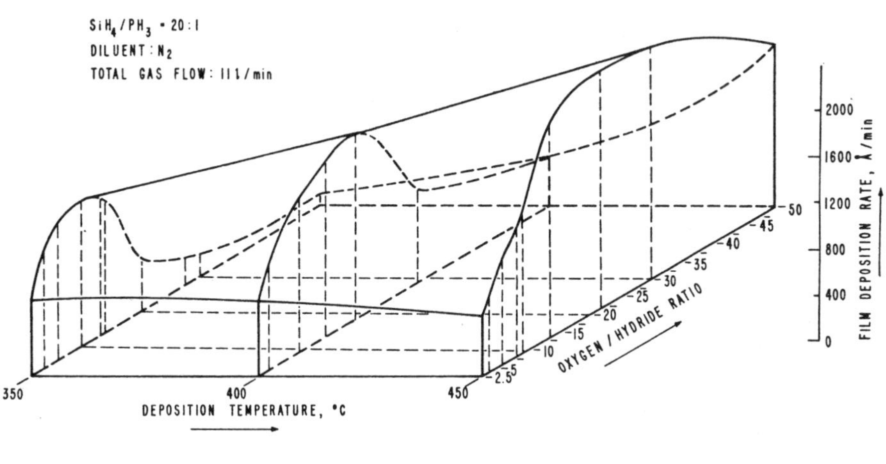

Figure 1: Deposition rate of PSG films.[2]

Varying the SiH_4/PH_3 ratio with other parameters held constant leads to different percentages of P_2O_5 in the SiO_2. For example, with 8% PH_3 in (SiH_4 + PH_3) at approximately 450°C, one obtains 6% P_2O_5 in (P_2O_5 + SiO_2). Too much phosphorus in the film can be deleterious, because it may result in a hygroscopic glass which can lead to metal corrosion problems.

The stress in such SiO_2 films when deposited on Si wafers in this process is always tensile. Typical stress values and their variation with deposition rate are shown in Figure 2. Note that higher deposition temperatures lead to higher stresses. Also, depositing the films with wet N_2 served to reduce the stresses found. The concern with stress arises when it is realized that thin films over uneven topography will end up cracking if they have tensile stresses built into them. Since one of the major applications of this film is as a final passivation layer over the completed integrated circuit, its tendency to crack is of concern. We will cover plasma silicon nitride films in Chapter 5, and we will note that one of the more desirable features of these films is that they tend to have compressive stresses.

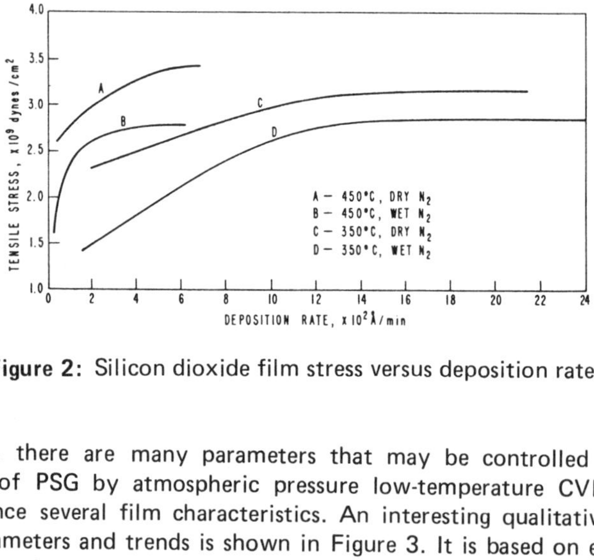

Figure 2: Silicon dioxide film stress versus deposition rate.[2]

Clearly, there are many parameters that may be controlled during the deposition of PSG by atmospheric pressure low-temperature CVD. They in turn influence several film characteristics. An interesting qualitative presentation of parameters and trends is shown in Figure 3. It is based on experiments done over a period of time.

The low-temperature depositions described in the present section can be used for either interlayer dielectrics or final passivation films. Their primary disadvantage is one of film quality, because the process is susceptible to gas-phase nucleation and incorporation of particles into the film.

3.2.2 Low-Pressure

In the mid-1970's, it was realized that low-pressure CVD processing could have significant advantages over atmospheric pressure systems. By reducing the pressure, it was found that the diffusion coefficient was sufficiently enhanced that deposition became surface controlled (see Chapter 2). In this case, wafers could be stacked closely and placed in a diffusion furnace to be processed

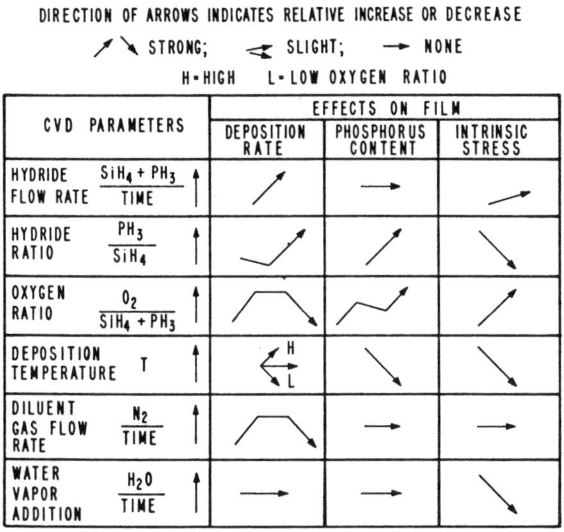

Figure 3: Effect of key parameters on CVD dioxide deposition.[2]

(hot-wall system). Since the temperature could be maintained very uniformly, deposition uniformity was excellent. Of equal importance, the large number of wafers that could be processed in each batch made the economics very favorable.[3] A typical hot tube reactor for low-pressure CVD is shown in Figure 4.

The major design problem with such a reactor involves finding conditions for which the deposition is not only uniform on each wafer, but is uniform from wafer to wafer. Of course, the deposition rate has to remain high to retain the throughput advantage.

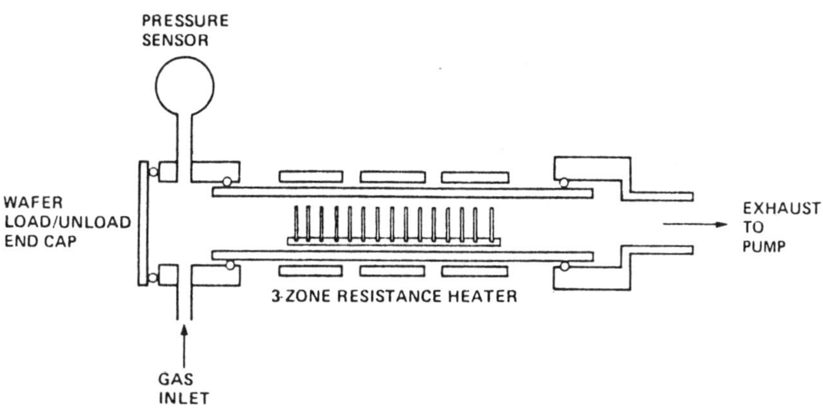

Figure 4: Low-pressure CVD reactor.[2]

First attempts were to run the process with SiH_4 and O_2. When this was done, films with good uniformity were deposited at 445°C and 1.0 Torr, as shown in Figure 5. However, the deposition rate was limited to 20 to 30 Å/min with a wafer spacing of ⅜". Attempts to push to higher deposition rates (by increasing SiH_4 and/or O_2 flow rates) led to hazy deposits which indicated gas phase nucleation. These deposition rates were too low to be competitive with the existing cold wall systems.

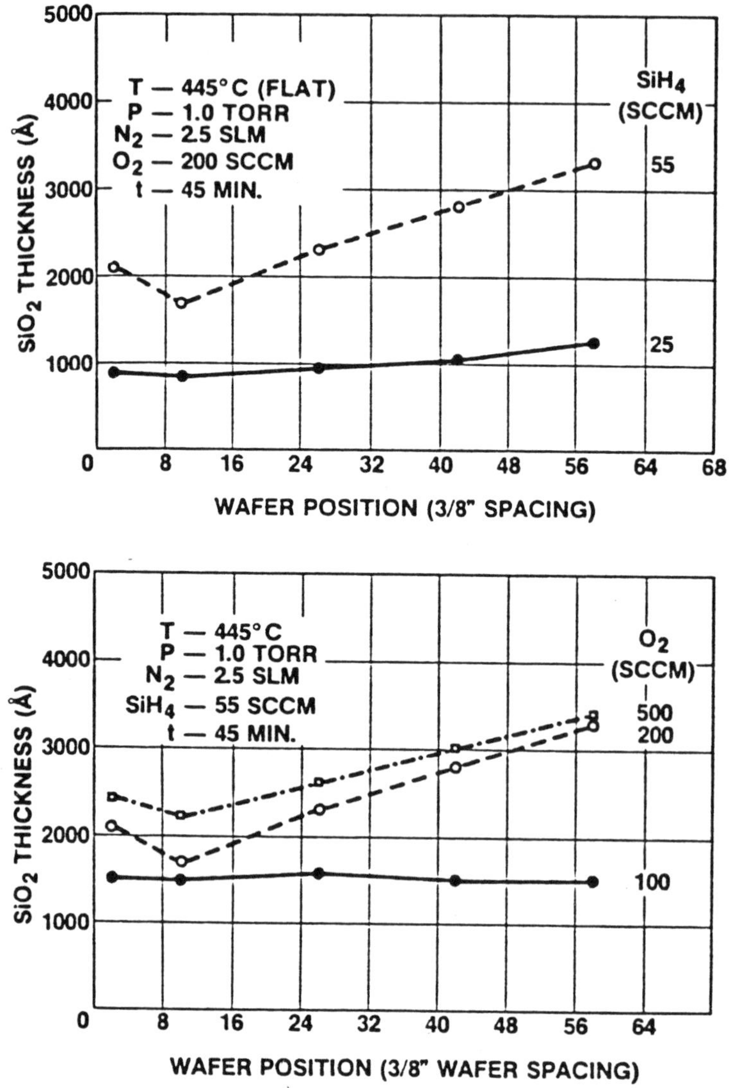

Figure 5: $SiH_4 + O_2$ reaction for SiO_2 films.[3]

Attention was then turned to a higher temperature, low-pressure process. Although such films would not be useful for final passivation, they could provide valuable interlayer dielectrics. At first, SiH_4 with N_2O was tried, where the reaction would proceed as follows:

$$SiH_4 + 2N_2O \longrightarrow SiO_2 + 2H_2 + 2N_2 \quad . \tag{1}$$

Good results were obtained at 840° to 860°C, in terms of uniformity, but again the deposition rate was low–less than 50 Å/min. Hazy films were again the result of attempts to push the deposition rate up.

Results with SiH_2Cl_2 and N_2O were more favorable. Here the reaction was assumed to be:

$$SiH_2Cl_2 + 2N_2O \longrightarrow SiO_2 + 2HCl + 2N_2 \quad . \tag{2}$$

At 900°C and a pressure of 0.8 Torr, good uniformity and clear films were achieved (see Figure 6) with deposition rates of 120 Å/min and wafer spacing of ⅜".

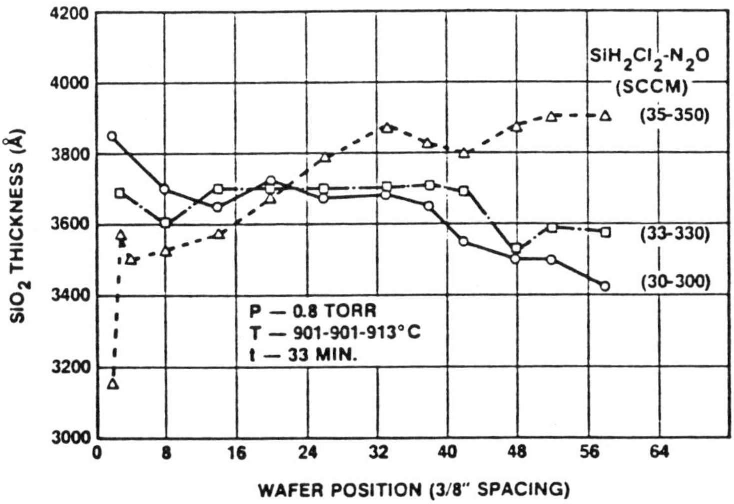

Figure 6: $SiH_2Cl_2 + N_2O$ reaction for SiO_2 films.[3]

Further investigation of the nature of these films[4] has shown that there is about 2.7% Cl remaining in the completed film. Upon exposing them to a thermal oxidation environment, it was discovered that the interface between the oxide and silicon was etched and film adhesion was lost–the films flaked off. Apparently, the thermal oxidation process released bound chlorine in the oxide which then diffused to the interface where it attacked the silicon.

Apparently, the low-pressure CVD process leaves some chlorine in the layer, while the same process at high pressure does not. Since we do not understand the kinetic pathways for the formation of this film, we cannot predict how to modify it. However, by trial and error it was discovered that small additions of O_2 to the N_2O feed gas reduced the chlorine content to zero.[4] Thickness uniformity was impacted by the O_2 addition, but a value of ± 7% was obtainable.

3.2.3 Reflow Phenomena

When phosphorus is added to SiO_2, in addition to gettering mobile alkali ions, it tends to reduce the intrinsic tensile stress in such films, thereby reducing their tendency to crack. Both functions are important when the film is used as a final passivation film for integrated circuits encapsulated in plastic. Phosphorus additions of 7 weight percent seem to be optimum in order to produce the above desirable film characteristics.

For interlayer dielectric, another consequence of the addition of phosphorus and/or boron is the ability to "reflow" the glass at lower temperatures. Lowering processing temperatures is a continuing objective in CVD processing.

When a dielectric film covers a polysilicon line, or a via is anisotropically etched down through an oxide film, we end up with nonplanar surfaces upon which one has to attempt to deposit aluminum by sputtering. Since sputtered aluminum does not produce a conformal coating, voids and/or thin regions can occur that will compromise device reliability. Heating the deposited SiO_2 to a high enough temperature will cause the glass to soften and flow, thereby rounding off sharp corners. Adding phosphorus and boron to the oxide lowers the temperature at which such reflow takes place. A sketch describing the reflow phenomena is shown in Figure 7.

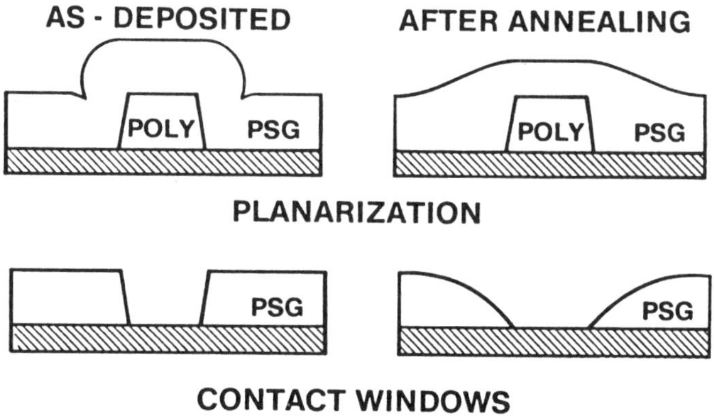

Figure 7: Reflow of PSG after heating.[5]

The top figures refer to the planarization of a deposit over a poly line. On the left is the as-deposited PSG. On the right, we see how much smoother the profile has become upon heat treatment. The lower figures refer to smoothing the corners of contact windows etched in PSG. On the left, we see the sharp corners produced by the typical anisotropic etch. On the right, these corners are substantially smoothed.

Experimental results for the reflow of PSG as a function of doping level for a 1000°C reflow in a N_2 ambient for 30 minutes is shown in Figure 8. The holes (windows) that had been etched in the 1-micron thick PSG had initially had sharp corners and vertical walls. Clearly, even the first sample with 7.34% doping received an acceptable rounding off.

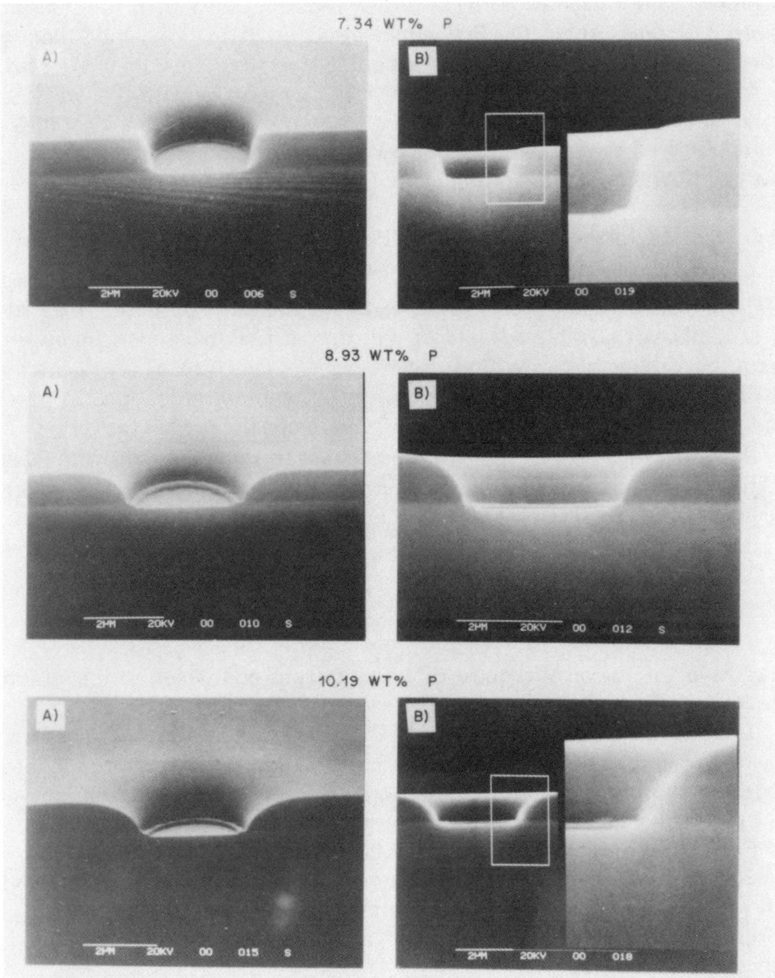

Figure 8: PSG reflow for different doping levels.[6] Reprinted by permission of the publisher, The Electrochemical Society, Inc.

Increasing the phosphorus content beyond the 6 to 8% level allows phosphoric acid to form in the presence of moisture. This acid can attack aluminum metalization, and can be the source of device reliability problems. It can also cause resist layers to lift, thereby terminating the processing. Recent research[5] has shown that adding B_2O_3 to the PSG by doping with B_2H_6 as well as PH_3 leads to a borophosphosilicate glass (BPSG) and the desired lowered viscosity at a given temperature.

The BPSG was deposited in a cold-wall CVD reactor operating at atmospheric pressure. Temperatures were 410° to 430°C, and the oxygen/hydride ratio was 20:1. Evaluating depositions, it was determined that the PH_3 flow had little if any effect on the B incorporation due to B_2H_6 flow. Therefore, the process was implemented by adjusting the B_2H_6 flow first, to establish the desired B doping. Then, the PH_3 could be adjusted to bring the P doping up to the proper level knowing that the B doping level would remain uneffected.

Reflow experiments established that films prepared with 3.4 wt % B and 4.5 wt % P had the same flow behavior, at a given temperature, as a PSG film with about 7 wt % P. Therefore, one can either add more phosphorus to further lower the flow temperature or maintain it as is and have a more reliable process.

3.2.4 Tetraethylorthosilicate (TEOS) Source

In addition to the SiH_4 or $SiCl_2H_2$ source gases that are used to deposit SiO_2, one can consider TEOS-$(C_2H_5O)_4Si$ for the same purpose. This material has a number of possible advantages. For one, it is a stable inert liquid, so the hazards of dealing with pyroforic silane are avoided. Also, it offers the potential of improved step coverage, which we will explore later. It does have the disadvantage of requiring a higher deposition temperature than the SiH_4 process.

As with earlier work, it would be desirable to dope the films with P and/or B. This has been tried[7] with liquid sources as well as phosphine and diborane. The liquid sources were:

Trimethylphosphate TMP-ate	$(CH_3O)_3PO$
Trimethylphosphite TMP-ite	$(CH_3O)_3P$
Trimethylborate TMB	$(CH_3O)_3B$

These have the major advantage of being safe, as contrasted to the very toxic phosphine and diborane.

The experimental setup in which TEOS was used for oxide depositions is shown in Figure 9. We see that the reactor is a low-pressure, hot-wall tube, so deposition uniformity on each wafer and from wafer to wafer will be the primary requirement as far as reactor operating conditions are concerned.

Experiments were carried out at 500 mTorr and temperatures from 620° to 680°C. The best results were obtained with phosphine and TMB doping, with added O_2 to increase the deposition rate. The amount of boron and phosphorus incorporated in the film could be varied by changing the TMB flow, as shown in Figure 10.

As noted earlier, a potential advantage of the TEOS process may be improved step coverage. To illustrate this phenomena, BPSG was deposited in a small, deep trench by both the TEOS and the SiH_4 processes. The resulting SEMs are shown in Figure 11.

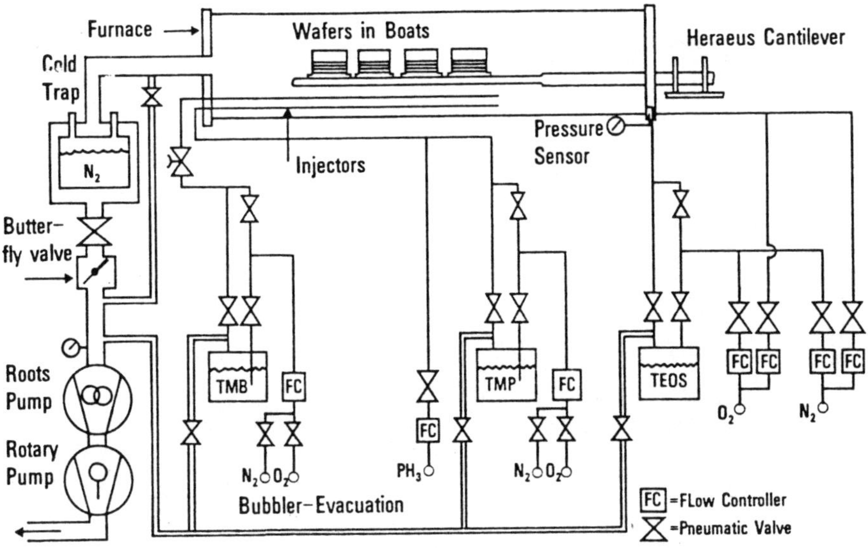

Figure 9: Schematic of setup for TEOS-BPSG depositions.[7]

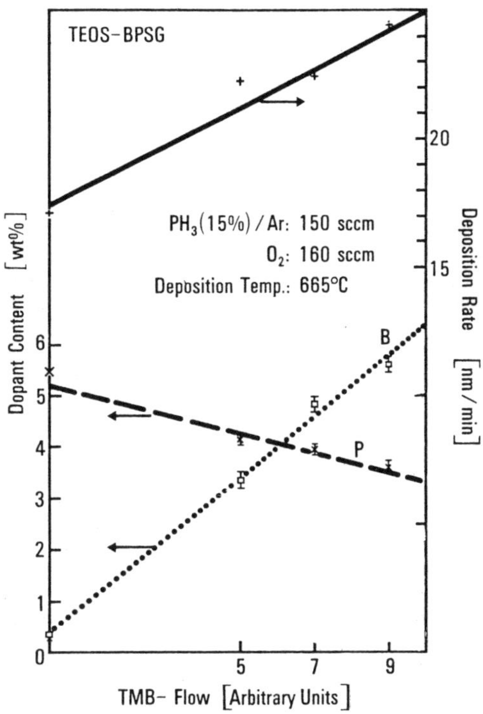

Figure 10: Influence of TMB flow on dopant content of TEOS-BPSG films.[7]

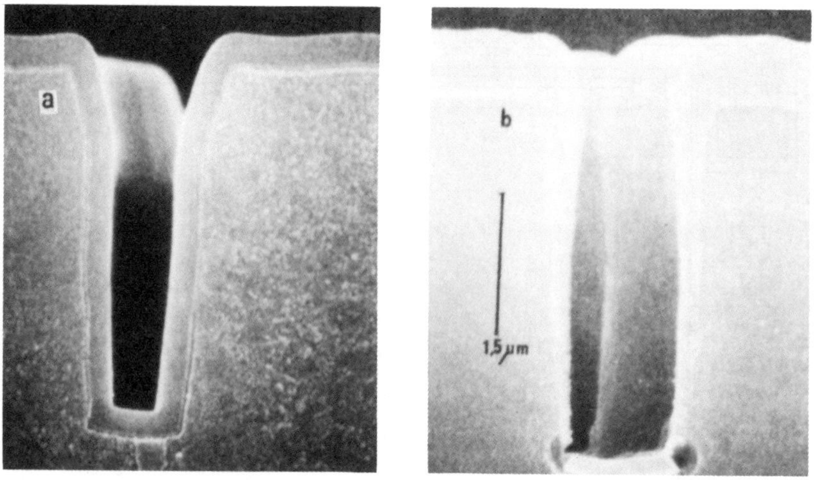

Figure 11: Step coverage for (a) TEOS-BPSG deposition at 640°C, and (b) SiH_4 process at 400°C.[7]

The explanation offered for this phenomena is that in the TEOS process, the molecule decomposes slowly after adsorbing on the surface, so there is more time for it to diffuse along the surface before breaking up and leaving SiO_2 behind. In the silane process, it is argued that the SiH_4 is so reactive that it decomposes as soon as it hits the surface, and therefore does not have time to diffuse down to the bottom of the trench before leaving SiO_2 behind. Also, there may be some temperature effect with molecules in the higher temperature process, (TEOS) being more mobile on the surface.

The films produced by these authors[7] exhibited good chemical stability with doping levels of 5 wt % B and 5 wt % P. Since these doping levels are higher than typically achieved in the SiH_4 process, reflow temperatures required can be lowered somewhat.

3.2.5 Diacetoxyditertiarybutoxysilane (DADBS) Source

The TEOS process just described has potential for use as an interlayer dielectric over metal layers that can withstand the high deposition temperature (i.e., doped polysilicon). If we want to cover aluminum, this process cannot be used. An alternative is to use a process where diacetoxyditertiarybutoxysilane (DADBS)-$(AcO)_2Si(OtBu)_2$ will decompose at temperatures between 450° and 600°C.[8]

Again, a low-pressure, hot-wall CVD reactor was used for the depositions. Pressures ranged from 300 to 900 mTorr. Phosphorus doping was carried out with trimethylphosphite. For these experiments, some depositions were carried out with oxygen addition. In this case, deposition rates were lowered. Combining trimethylphosphite with oxygen, on the other hand, increased deposition rates.

In comparison to the TEOS process, it is claimed that the step coverage is perfect. For example, taking the top surface adjacent to a trench as unity thickness, a trench with aspect ratio of 2:1 when coated with a 700°C TEOS process produces an 85% sidewall thickness and a 70% bottom thickness. It is claimed that a 450° to 500°C DADBS film is 100% on both sidewall and bottom.

3.3 SILICON NITRIDE

Again, Si_3N_4 can be deposited by atmospheric-pressure CVD, but the economics are more favorable to the LPCVD approach. Films from SiH_4 and NH_3 can yield stoichiometric Si_3N_4. More uniform and higher deposition rate films are created, however, from SiH_2Cl_2 and NH_3 in a reactor such as the one shown in Figure 4.[3] Under conditions such as 750°C, 500 mTorr, and NH_3/SiH_2Cl_2 = 10:1 flow ratios, excellent thickness uniformities are obtained. Within a wafer, uniformities are ± 3%, and from wafer to wafer they are ± 1.5%. Deposition rates are 37 Å/min.

A very detailed analysis of the chemical nature of thin (50 to 500 Å) LPCVD Si_3N_4 films has been carried out.[9] These authors show that Si_3N_4, by the above LPCVD process when deposited on Si, exhibits a 15 to 20% Å SiO_2 layer between the Si and Si_3N_4. It is felt that this layer originates from the native oxide on the Si, even after careful cleaning. Also, they detected 0.4% chlorine in the film originating from the SiH_2Cl_2, although the Si_3N_4 was indeed stoichiometric.

There did not appear to be any gross evidence of hydrogen in the films, but none of the analytical techniques used by these authors to evaluate the film composition were sensitive to this impurity.

3.4 POLYSILICON

Up to this point, all of the films we have considered (SiO_2, Si_3N_4) were deposited under conditions such that they were amorphous. The only defects of interest were particles from the gas phase that might be incorporated into the growing film, or pinholes. Low-pressure CVD has reduced the incidence of particles, and thicker films can minimize the presence of pinholes.

When we consider silicon films, on the other hand, the nature of the solid deposit is crucial to the behavior of the film. Depending on deposition conditions, we can deposit amorphous, polycrystalline, or single crystal films. As was noted in Chapter 1, the morphology of polycrystalline films can be complex. In the present section, we will review some aspects of polysilicon (poly) thin films deposited by CVD. The final section of this chapter will be devoted to epitaxial silicon thin films.

3.4.1 Deposition Behavior

First, let us consider some of the factors that arise when we form CVD poly, and then when we try to dope it in-situ. Although CVD poly can be deposited in a cold-wall atmospheric-pressure reactor, economic considerations are such that today it is only deposited in a low-pressure hot-tube furnace. The basic chemistry used is the pyrolysis of SiH_4, usually with a diluent gas such as N_2, He, H_2, etc. Other chemistries have been tried, but none has yielded as uniform depositions at the high deposition rates as the SiH_4 process. The influence of temperature and pressure on deposition over many wafers in a tube furnace[3] is shown in Figure 12. Clearly, uniform wafer-to-wafer depositions can be achieved by modifying the reactor temperature along the reactor axis. Although deposition rates are higher at the higher pressures (700 mTorr), the within-wafer uniformity is not as good, so depositions tend to be done at 500 mTorr. When pure SiH_4 is used, extreme sensitivity to SiH_4 flow is found, as illustrated in Figure 13. In addition, the deposition rate is reduced from 200 Å/min with 23% SiH_4 to 125 Å/min with 100% SiH_4.

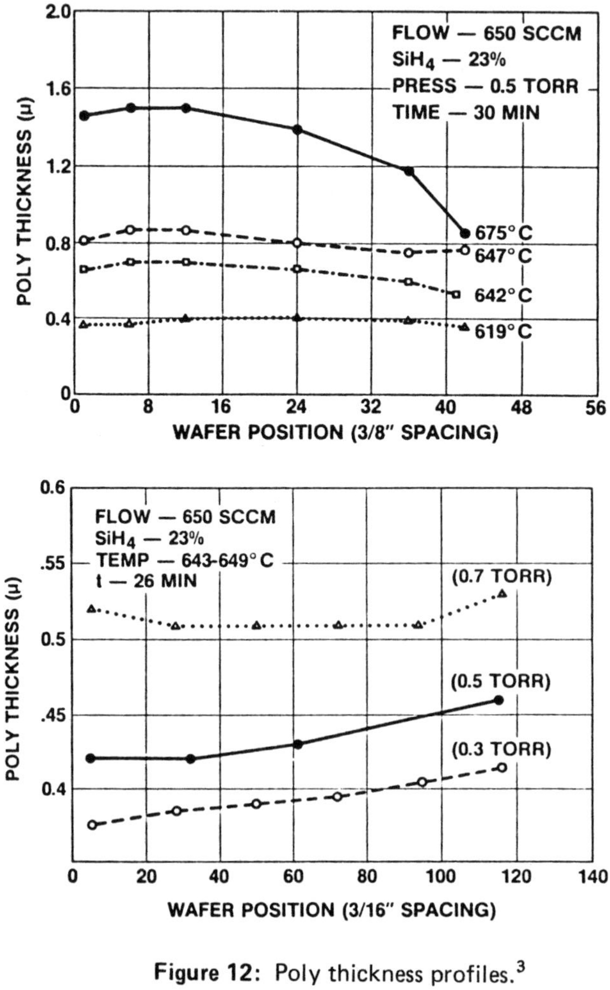

Figure 12: Poly thickness profiles.[3]

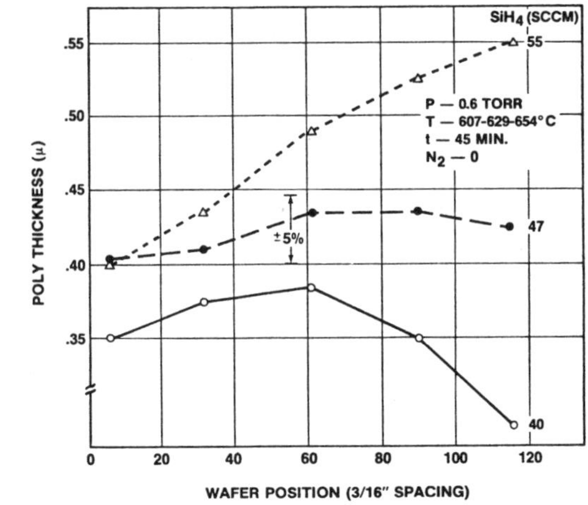

Figure 13: Poly thickness profiles for 100% SiH_4.[3]

So far, we have ignored the primary reason poly films are used in integrated circuits. Heavily-doped poly is used as a gate electrode, and the electrical conductivity of this material is of prime importance. Therefore, we have to inquire into the feasibility of doping poly as it is being deposited by CVD.

The obvious approach to this problem would be to deposit from a SiH_4 + PH_3 mixture, in the hopes that a sufficient quantity of P dopant could be incorporated into the poly. Many attempts to do this have been unsuccessful. For example, depositions at 623°C and 100 mTorr were carried out with 30 sccm of SiH_4 and 0.75 sccm of PH_3.[10] Without the PH_3, deposition across a single wafer was very uniform (see Figure 14). Adding the small quantity of PH_3 reduced the deposition rate at the wafer center by 20:1, and yielded a nonuniform film.

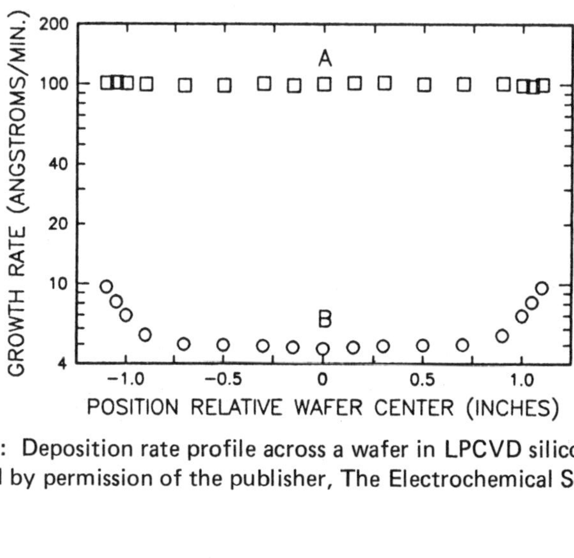

Figure 14: Deposition rate profile across a wafer in LPCVD silicon deposition.[10] Reprinted by permission of the publisher, The Electrochemical Society, Inc.

There are two issues that have to be resolved when reviewing these results. One—why is the deposition rate reduced by such a large factor? Second—why is the deposition nonuniform now, when it was uniform without the PH_3? The explanation proposed[10] for the first question is that PH_3 preferentially adsorbs on the silicon surface, preventing SiH_4 from adsorbing and subsequently decomposing to Si and H_2.

The explanation for the second question is somewhat more involved. It is suggested that deposition occurs both as a result of SiH_4 decomposition on the surface (heterogeneous reaction), and SiH_4 decomposition in the gas phase (homogeneous reaction) to SiH_2 and H_2 and subsequent deposition by SiH_2 reaching the surface. Because SiH_2 is a free radical, it should react at the surface with unit probability. This would have the effect of depleting the SiH_2 from the gas phase, and return the process to one controlled by diffusion rather than surface kinetics. Therefore, a thinner poly film at the wafer center is a reasonable expectation.

Because of the above-mentioned difficulties in trying to dope poly with phosphorus in situ, such films have traditionally been deposited undoped. Doping can then be accomplished by ion implantation or diffusion.

3.4.2 Electrical Resistivity of Doped Films

Regardless of the method of doping the LPCVD poly film (in situ, implant or diffusion), the important fact is that the resistivity of the film is higher than a doped epi film would be.[11] Resistivity measurements are shown in Figure 15 for both phosphorsus and boron doping. For low doping concentrations, the resistivity can be five orders of magnitude higher. Even at higher concentrations, we still see an order of magnitude greater resistivity.

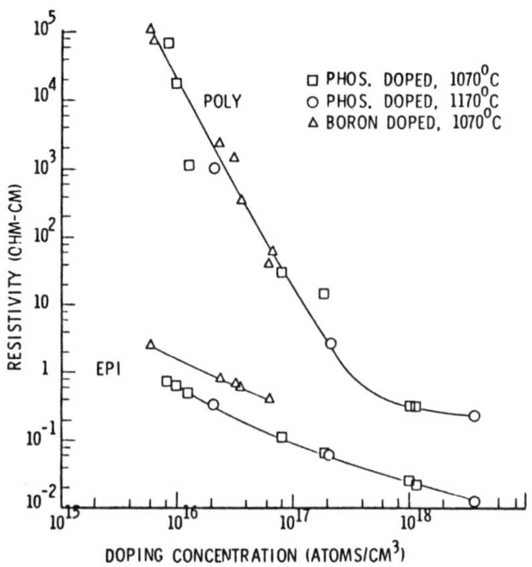

Figure 15: Resistivity versus doping concentration.[11] Reprinted by permission of the publisher, The Electrochemical Society, Inc.

It is fairly obvious that this phenomena must be related to the fact that the poly is made up of many small grains with their attendent grain boundaries. Each grain is a single crystal and should behave as the epi films do. Therefore, the cause of the increased resistivity must involve the grain boundaries.

There are two proposals as to why grain boundaries should effect the electrical properties of doped poly films. One is that dopant atoms segregate at the grain boundaries where they are electrically inactive,[12] thereby reducing the dopant concentration in the grains themselves. For lightly-doped films, most of the dopant segregates at the grain boundaries. At higher dopant levels, the grain boundaries become saturated and the remaining dopant goes directly into the grains. Thus, heavily-doped poly films approach doped epi films as far as resistivity is concerned.

The second argument states that since the atoms in the grain boundaries are disordered, incomplete atomic bonding could lead to the formation of trapping states. Such trapping states then could immobilize carriers, and so reduce the number of free carriers available to conduct electricity.[13] Once the mobile carriers are trapped, the grain boundaries become electrically charged, creating a potential barrier to the flow of carriers from one crystal to another.

Undoubtedly, both explanations play some role, since each has some experimental verification to support it. It will depend on the type of dopant, the extent of dopant, and grain size. As noted earlier in Chapter 2, poly grains will grow larger when the film is doped, as well as annealed at a high temperature for a reasonable length of time.

In support of a dual mechanism, recent experimental evidence[14] has shown that a significant fraction of arsenic and phosphorus dopants may appear at grain boundaries, especially for moderate dopant concentrations. The fraction segregating to the grain boundaries is found to be inversely proportional to the size of the grains. For boron-doped films, no evidence was found of dopant segregation at the grain boundaries.

3.5 EPITAXIAL SILICON

In the previous section, we discussed the CVD of silicon thin films. For the pressures and temperatures at which those depositions were carried out, the films were polycrystalline. If the depositions had been carried out at higher temperatures, single-crystal (epitaxial) films would have been possible. In this section, we will discuss some of the factors that govern the growth of epi silicon films.

Early attempts to build integrated circuits on single-crystal wafers cut from single-crystal boules met with many difficulties due to the high concentration of impurities present. A solution to this practical problem was found by growing epitaxial (epi) silicon by CVD on top of the impure wafers. Since the feedstock gases used in the epi process (dichlorosilanes mixed with hydrogen) could be highly purified, the resulting epi films were much purer than the underlying substrate. It was in this very pure epi film (so-called "device quality" silicon) that high-quality integrated circuits could be built. The single-crystal wafer became, in effect, just a mechanical holder for the epi film (of course, one with the correct crystal structure).

In the early integrated circuit developments, the epi layers were generally quite thick. A typical layer could be 20 microns thick. Therefore, high deposition rates were essential to the economic viability of the equipment. The highest deposition rates are possible with atmospheric pressure reactors at high temperatures. As discussed in Chapter 1, the higher temperature depositions tend to be diffusion controlled, and this ruled out hot tube systems with many wafers in a single batch. Accordingly, the industry standard has been the cold-wall batch reactor (barrel type) run at atmospheric pressure. Some systems are being operated at 80 Torr, but this is still in the diffusion-controlled regime.

Some of the specifics of the epi silicon CVD process will be covered in the balance of this chapter.

3.5.1 The CVD Process for Epi Silicon

The reactions that have been used to create epi silicon films commercially involve the H_2 reduction of the chlorosilanes. As we learned earlier in Chapter 1 in studying the equilibrium behavior of the H-Cl-Si system, we can deposit solid silicon from $SiCl_4 + H_2$, $SiCl_3H$, or $SiCl_2H_2$. Also, H_2 can be added to the latter two, if desired. Obviously, silicon will also deposit from SiH_4. The deposition rates of Si as a function of temperature, at atmospheric pressure, from the CVD of the above source gases are shown in Figure 16.

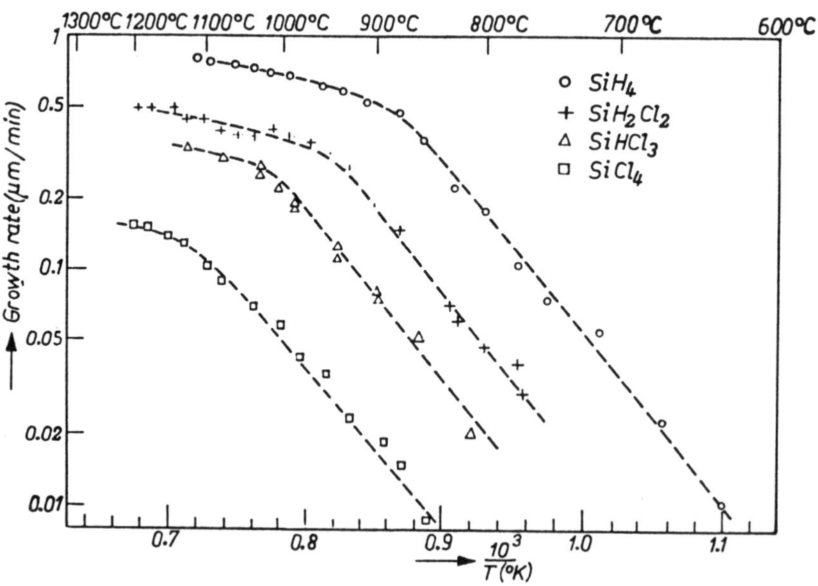

Figure 16: Silicon growth rates as a function of temperature. All reactants ~0.1 mol % in H_2.[15]

Several features of these curves are significant. At low temperatures, the reactions are considered to be surface controlled, so there is an overabundance of reactant near the surface and the rate of deposition varies rapidly with surface temperature. At higher temperatures, the deposition rate appears almost independent of temperature, as would be expected for a diffusion-controlled deposition.

At the lower temperatures, the films are polycrystalline. Epitaxial films are produced for temperatures at or above values at the knee of each curve. Therefore, epi can be grown from SiH_4 at temperatures as low as 800°C. On the other hand, epi films from $SiCl_4$ require deposition temperatures above 1100°C.

In growing epi films, the most commonly used reactant is $SiCl_4$. It is inexpensive and easily available. Another reason for its popularity is that it leaves relatively little deposit on the cold walls of the reactor bell jar, so that cleaning is less of a problem. Also, when deposits are done at temperatures between 1100° and 1300°C, film quality is excellent in terms of crystallographic defects.

Where thicker films are needed, $SiHCl_3$ is often used because of its higher deposition rate. In other respects, it is similar to $SiCl_4$. When lower deposition rates become critical, $SiCl_2H_2$ is being used.

Finally, epi films deposited from SiH_4 at low temperatures (~1000°C) are interesting, but difficulties due to the heavier Si deposits on the cold reactor walls has limited interest in this approach.

The typical epi silicon reactor operates in the diffusion-controlled regime at high rates of deposition. The behavior of such a reactor is governed by the fluid dynamics of multicomponent gases. The gas phase reactions discussed in Chapter 1 are generally neglected.

In principle, epi reactors that operate in the diffusion-controlled regime could be designed by solving the partial differential equations governing the fluid dynamics[16,17] so that deposition rates could be predicted. In fact, such a procedure is generally not followed, since experimental evaluation of the flow behavior seems to be preferred.[18]

3.5.2 Surface Effects

Using the proper CVD process for Si deposition in a system which has the fluid mechanics properly arranged is not sufficient to produce quality epi Si films. Assuming we are hoping to grow on a single-crystal substrate, this substrate surface must be properly prepared. It must have "atomic" steps on the surface to provide nucleation sites. Such atomic steps are obtained by cutting the substrate several degrees off the normal to the boule growth axis.

Secondly, the wafer must be very "clean." Even a clean substrate will have 20 to 50 Å layer of native oxide on it, and/or some carbon, and this will be enough to impede nucleation and give rise to many defects.[15] After wafers are cleaned and inserted into the reactor, there is still the oxide layer to be removed as well as possibly some carbon on the surface. The traditional way of dealing with this phenomena is to operate a high-temperature HCl (1200°C) etch before attempting depositions. This etches away the native oxide, and any carbon on the surface diffuses into the bulk at this temperature.

It is also thought that the success of the chlorosilane + H_2 process, in producing high-quality epi Si films, is related to the HCl produced in the reaction. It is thought that the process is close to equilibrium, and that there is significant etching by HCl going on while Si is being deposited.

3.5.3 Defects

Even when epi silicon films are successfully grown, defects in the film can still be observed. In a commercial reactor, it is never possible to drive the concentration of such defects to zero. Specifications are usually defined as #/cm^2 allowable.

The common defects can be seen with optical microscopes, and they become more clearly visible after suitable etches.[20] The most common defects are stacking faults and spikes (see Figure 17). These can be caused by local surface imperfections as well as surface particulates. Another defect frequently occurring is the slip lines shown in Figure 18.

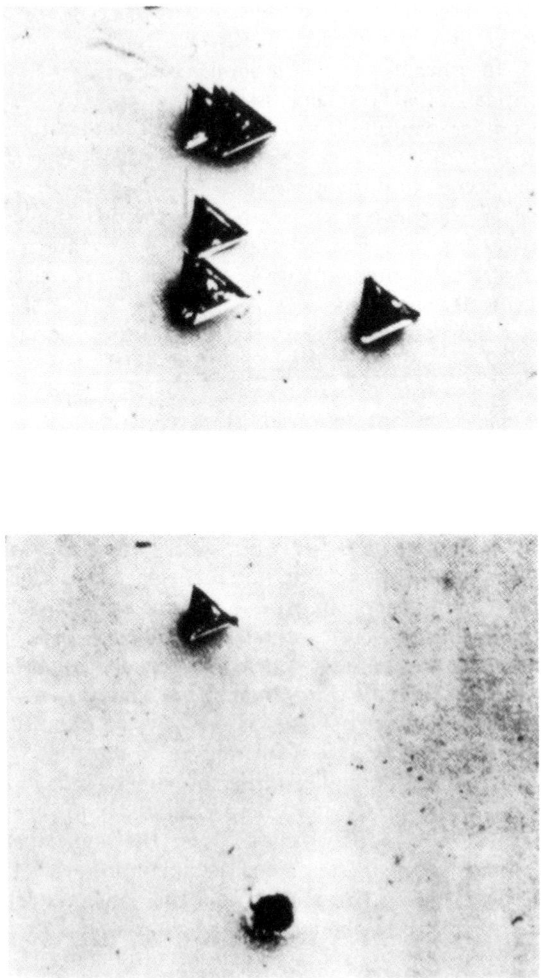

Figure 17: Stacking faults and spikes.[19]

Figure 18: Slip lines with stacking faults.[19]

If the high-temperature (1200°C) etch is used and a high-temperature deposition as well, stacking faults and spikes tend to be minimized. Unfortunately, this is when slip becomes a real problem. Slip occurs as parts of the single crystal move relative to each other along crystallographic planes, due to high thermal stresses. They generally occur at the outer edge of the wafer where it is stressed. For example, a common location for such slip defects are the points at which the wafer edge rests on the susceptor.

Obviously, if there are a huge number of such defects, it will be impossible to build quality devices on such wafers. Even when there are a few such defects, they can be very harmful because they seem to attract metallic impurities. Thus, what started out as mechanical defects gives rise to metallic precipitation defects which are much more damaging to circuit operation.[21]

3.5.4 Autodoping

In the fabrication of integrated circuits, heavily-doped islands are created in the bare substrate surface. This surface is then covered with a lightly-doped epi film. The objective is to achieve a sharp junction between the heavily- and lightly-doped regions. If the epi layer above the doped region is contaminated with dopant, this is called vertical autodoping. If the epilayer to the side of the buried layer is contaminated, this is referred to as lateral autodoping.

Autodoping of epi films can be explained by two mechanisms. For one, dopant could diffuse (solid state diffusion) from the buried layer to the epi film during its formation. Second, the dopant from the buried layer can vaporize, enter the reactor gas flow, and be incorporated as the surface reaction proceeds. The concensus seems to be that the latter effect is the predominant one. In fact, it is well known that coating the back of the wafer with oxide reduces the autodoping, and this can only relate to gas phase transport.

Reactor operating conditions also play a role, since it is well known that arsenic autodoping is reduced when the reactor is operated at reduced pressures (i.e., 80 Torr).[22] Because of the ability to reduce arsenic autodoping drastically at low pressure, this has become an important commercial process.

For epi depositions with arsenic buried layers, we can see the influence of pressure on dopant profile for the $SiCl_4$ process in Figure 19 and for the SiH_2Cl_2 process in Figure 20. In both cases, as the pressure is reduced, the width of the transition region is less. Measurements were made by SIMS. The heavily-doped buried layer substrate is shown on the right-hand side of these figures, and the epi film is on the left.

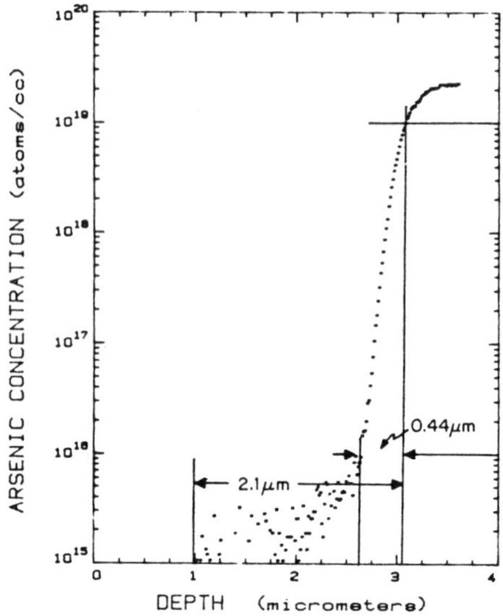

Atmospheric Pressure

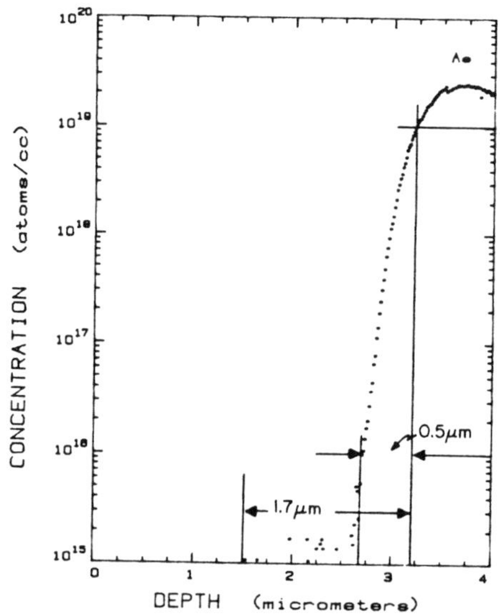

100 Torr

Figure 19: Deposition from $SiCl_4$ at 1150°C.[24]

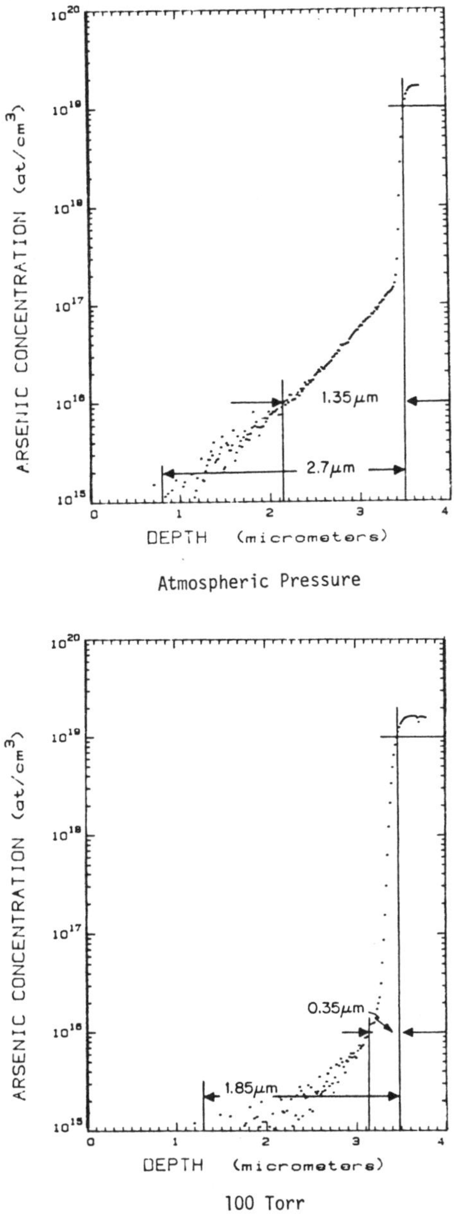

Figure 20: Deposition from SiH_2Cl_2 at 1080°C.[24]

Another interesting feature that can be seen from Figures 19 and 20 is the thinner transition region achieved with $SiCl_4$ as compared to $SiCl_2H_2$. It appears that the thinnest transition regions occur with these molecules that have the greatest chlorine concentrations. It is thought that the HCl generated in the reaction can etch the growing film at the impurity lattice sites, and thereby minimize the inclusion of dopant.

Finally, we should note that although arsenic autodoping can be substantially reduced by low-pressure operation, experimental evidence suggests that boron autodoping is not influenced by pressure.[23]

3.5.5 Pattern Shift

The first step in the fabrication of modern bipolar integrated circuits is the creation of heavily-doped islands in the silicon substrate within which to fabricate the circuits. In the process of creating these "buried layers," the area they occupy has a surface that lies somewhat below the surface of the undoped substrate. The typical appearance of such a buried layer is illustrated in Figure 21.

Before Epi

Figure 21: Buried layer test patterns.[24]

Pattern shift is the phenomena that occurs when one attempts to grow an epilayer on such a nonuniform surface. The epi film has different growth rates on different crystal faces. For example, deposition is fastest on (112) and slowest on (111). Pattern shift is illustrated in Figure 22 for $SiCl_4$ and Figure 23 for SiH_2Cl_2. Here we can clearly see that reduced pressure has minimized this problem in both cases.

Atmospheric Pressure
t = 11.2 μm

100 Torr
t = 13.8 μm

Figure 22: Pattern shift in $SiCl_4$ deposition.[24]

Atmospheric Pressure
t = 12.5 µm

100 Torr
t = 17.5 µm

Figure 23: Pattern shift in SiH_2Cl_2 deposition.[24]

3.5.6 Low-Temperature Epi Silicon

As device geometries become smaller and smaller, the epi layers also scale down and become thinner. The thinner the epi layer, the more serious the autodoping problem becomes. Very sharp transitions are needed, and cannot be obtained readily at normal temperatures (>1000°C). Therefore, there is great interest in finding a low-temperature epi process that produces good quality epi layers.

A number of researchers have grown epi films at temperatures as low as 800°C.[25] However, in all cases, a high temperature (1040° to 1180°C) clean step was first needed.

The high-temperature cleaning or etching step can be avoided by using a plasma to etch at low temperature.[26,27] Both H_2 and Ar plasmas were tried. When H_2 is used, there should be both a chemical etch effect as well as sputtering due to ion bombardment. With Ar, the principal effect should be physical sputtering of impurities and native oxide. It is interesting to note that the key feature of the argon clean procedure is that deposition gas flow must overlap the etching process.[27] If this is not done, the Si grows a new native oxide layer in 1 second.

Using these techniques, epi films have been deposited at temperatures as low as 650°C. At these temperatures, there is no autodoping problem. However, whether or not these are useful films for devices is not yet proven. Also, it remains to be seen whether a reactor based on these techniques will be economically viable.

Another approach to this problem involves heating the wafer at 750°F at very low pressures ($<10^{-10}$ Torr) prior to deposition.[28] This has the effect of removing the native oxide by evaporation of SiO. Depositions were achieved in the temperature range of 750° to 850°C in $SiH_4 + H_2$. Since the authors were developing a hot-wall system with many wafers stacked close to each other, the deposition was carried out at 2 mTorr. Deposition rates of 20 to 45 Å/min were achieved. As expected, dopant transition widths were very narrow, several hundred angstroms. Again, device studies on such a system have not yet been done.

REFERENCES

1. Kern, W., Chemical Methods of Film Deposition, in *Thin Film Processes,* eds. J.L. Vossen and W. Kern, Academic Press, NY (1978).
2. Kern, W. and Rosler, R.S., Advances in deposition processes for passivation films. *J. Vac. Sci. Technol.* 14:1082 (1977).
3. Rosler, R.S., Low pressure CVD production processes for poly, nitride, and oxide. *Solid State Technol.* 20(4):63 (1977).
4. Kemlage, B.M., Film integrity of high-temperature LPCVD-SiO_2 in *Chemical Vapor Deposition,* Eighth International Conference on Chemical Vapor Deposition, eds. J.M. Blocker, Jr., G.E. Vuillard and G. Wahl [(Electrochemical Society, Pennington, NJ (1981)], pg. 418.
5. Ramiller, C.L., and Yau, L., Borophosphosilicate glass for low temperature reflow. *Semicon. West Techn. Proc.* 5:29 (1982).
6. Levy, R.A., Vincent, S.M., and McGahan, T.E., Evaluation of the phosphorus concentration and its effect on viscous flow and reflow in phosphosilicate glass. *J. Electrochem. Soc.* 132:1472 (1985).
7. Becker, F.S., Pawlik, D., Schafer, H. and Standigl, G., Process and film characterization of low pressure tetraethylorthosilicateborophosphosilicate glass. *J. Vac. Sci. Technol.* B4(3):732 (1986).
8. Smolinsky, G., The low pressure chemical vapor deposition of silicon oxide films in the temperature range 450° to 600°C from a new source: diacetoxyditertiarybutoxysilane, in Proceedings of the 1986 Symposium on VLSI Technology, San Diego, May 1986 (IEEE Catalog #86CH2318-4).
9. Habraken, F.H.P.M., Kuiper, A.E.T., Oostrom, A.V., and Tamminga, Y., Characterization of low-pressure chemical vapor deposited and thermally grown silicon nitride films. *J. Appl. Phys.* 53(1):404 (1982).
10. Meyerson, B.S., and Olbricht, W., Phosphorus-doped polycrystalline silicon via LPCVD; I. process characterization. *J. Electrochem. Soc.* 131:2361 (1984).
11. Fripp, A.L., and Slack, L.H., Resistivity of doped polycrystalline silicon films. *J. Electrochem. Soc.* 120:145 (1973).
12. Cowher, M.E., and Sedgwick, T.O., Chemical vapor deposited polycrystalline silicon. *J. Electrochem. Soc.* 119:1565 (1972).
13. Kamins, T.I., Hall mobility in chemically deposited polycrystalline silicon. *J. Appl. Phys.* 42:4357 (1971).
14. Mandurah, M.M., Saraswat, K.C., and Helms, R.C., Dopant segregation in polycrystalline silicon. *J. Appl. Phys.* 51(11):5755 (1980).
15. Bloem, J., and Giling, L.J., Mechanisms of the Chemical Vapor Deposition of Silicon, in *Current Topics in Materials Science, Vol. I,* ed. Kaldis, E., [(North-Holland Publishing (1978)].
16. Klingman, K.J., and Lee, H.H., Design of epitaxial CVD reactors, I. Theoretical relationships for mass and heat transfer, *J. Crys. Growth* 72:670 (1985).
17. Toor, I.A., and Lee, H.H., Design of epitaxial CVD reactors, II. Design considerations and alternatives. *J. Crys. Growth* 72:679 (1985).
18. Corboy, J.F., and Pagliaro, R., Jr., An investigation of the factors that influence the deposit/etch balance in a radiant-heated silicon epitaxial reactor, *RCA Review* 44:231 (1983).

19. Atherton, R.W., Fundamentals of silicon epitaxy. *Semiconductor International* (Nov. 1981), p. 117.
20. Jenkins, M.W., A new preferential etch for defects in silicon crystals. *J. Electrochem. Soc.* 124:757 (1977).
21. Werkhoven, C.J., Source transport and precipitation of metallic impurities in Si epitaxy. in *Aggregation Phenomena of Point Defects in Silicon,* eds. Sirth, E. and Goorissen, J. [(Electrochemical Society Pennington, NJ (1983)], Vol. 83-4, p. 144.
22. Ogirima, M., Saida, H., Suzuki, J. and Maki, J., Low pressure silicon epitaxy. *J. Electrochem. Soc.* 124:903 (1977).
23. Kulkarni, S.B., and Kozul, A.A., Boron autodoping in reduced-pressure epitaxy. The Electrochemical Society Extended Abstracts [(Electrochemical Society, Pennington, NJ (1980)], Abstract No. 540, p. 1351.
24. Cullen, G.W., Corboy, J.F., and Metzl, R., Epitaxial reactor systems: Characteristics, operation, and epitaxy costs. *RCA Review* 44:187 (1983).
25. Richman, D., Chiang, Y.S., and Robinson, P.H., Low temperature vapor growth of homoepitaxial silicon. *RCA Review* 31:613 (1970).
26. Townsend, W.G. and Uddin, M.E., Epitaxial growth of silicon from SiH_4 in the temperature range 800° to 1150°C. *Solid State Electronics* 16:39 (1973).
27. Donahue, T.J., Burger, W.R. and Reif, R., Low temperature silicon epitaxy using low-pressure chemical vapor deposition with and without plasma enhancement. *Appl. Phys. Lett.* 44:346 (1984).
28. Meyerson, B.S., Gannin, E. and Smith, D.A., Low temperature silicon epitaxy by hot wall ultra high vacuum/low pressure chemical vapor deposition techniques. Electrochem. Soc. Fall Mtg., Oct. 1985, Extended Abstracts 85-2, pg. 401.

4

Thermal CVD of Metallic Conductors

4.1 INTRODUCTION

As before, we observe that there are many metallic conducting films that can be deposited by CVD.[1] It is not our intention to catalogue all of these. Rather, we will restrict our attention to those films either in use in integrated circuit manufacture, or that have good potential for such use.

In contrast to the films described in the last chapter, the ones to be discussed in this chapter have only become of interest recently. Up to the present, the integrated circuit gate electrodes have been fabricated from LPCVD polysilicon, which is heavily doped with phosphorus in a separate step (either by diffusion or ion implantation). Such heavily doped polysilicon can have resistivities as low as 500 $\mu\Omega$-cm, so it behaves as a conductor, although not a very good one. Its compatibility with standard processing steps, however, make it a very attractive gate material.

The final metallization of the standard single-layer metal conductor circuits has been provided by sputtered aluminum. As required, the sputtered Al can be doped with Si to minimize spiking of Al into the Si that it must contact. It can also be doped with copper to minimize electromigration effects.

In recent years, VLSI requirements have led to closely spaced long interconnection lines with smaller cross sections.[2] The ensuing RC time delay can limit the speed with which circuits can be operated. Also, the power consumption due to high resistance can be appreciable and heat the circuits more than permitted. Therefore, the doped poly available is becoming inadequate for the new generation of circuits. This has led to the development of refractory metal silicide films because of their high-temperature processing capability. Initially, they were deposited by evaporation or sputtering. These are WSi_2, $MoSi_2$, $TaSi_2$ and $TiSi_2$.

The first problem occurs with the gate electrode. The solution that has been developed has been to create a "polycide" structure. Here, a thin layer of phosphorus-doped poly is deposited and then a conducting layer of silicide is deposited on top of the first layer. The combination is a much better conductor than the doped poly. Additional details of this polycide film will be covered later. The properties of the silicide films have been reviewed by Murarka.[3] We will restrict our discussion to these films when they are deposited by CVD.

Again, as VLSI requirements become more demanding, multilevel conductor circuits are being developed. The final metallization layer can be aluminum, since there are no additional processing steps that require temperatures above 350°C. However, if we wish to use a second conductor level between the gate electrode and the final metallization, then aluminum is no longer acceptable. It melts at about 660°C, should not be heated above 500°C, and there would be additional processing steps well above these temperatures. Some integrated circuit developers have used two layers of polysilicon in this application, and others have tried to develop low-temperature processing techniques for dielectric deposition to permit two aluminum levels. Both approaches have severe shortcomings, so CVD of refractory metals has some attraction. The resistivity of the refractory metals or silicides are not as good as aluminum, but for tungsten or molybdenum, it can be within a factor of two, a large improvement over doped poly. So, in addition to the refractory metal silicides, there is much interest in refractory metals, and these will be discussed later.

As a point of interest, approximate values of the thin film resistivities of these materials are tabulated in Table 1.

Table 1: Resistivities of Representative Thin Film Metallic Conductors

Material	Resistivity ($\mu\Omega$-cm)
WSi_2	50
$MoSi_2$	100
$TaSi_2$	50
$TiSi_2$	25
Mo	8
W	9
Al	4
Doped PolySi	500

An excellent presentation showing the future direction of gate/interconnect materials is shown in Figure 1.[4] Since 1-MB DRAMs are now appearing on the market, the pressure to move to the newer metallic conductors is strong.

As a final point, we note that as device dimensions shrink, it becomes increasingly difficult to obtain good conformal coverage over steps and trenches when line-of-sight techniques are used (i.e., evaporation or sputtering). In general, CVD offers excellent conformal coverage, so that has provided a further push toward CVD metals. In fact, poor step coverage with the traditional sputtered aluminum has led to an interest in CVD aluminum. We will conclude our chapter with a review of work in this area.

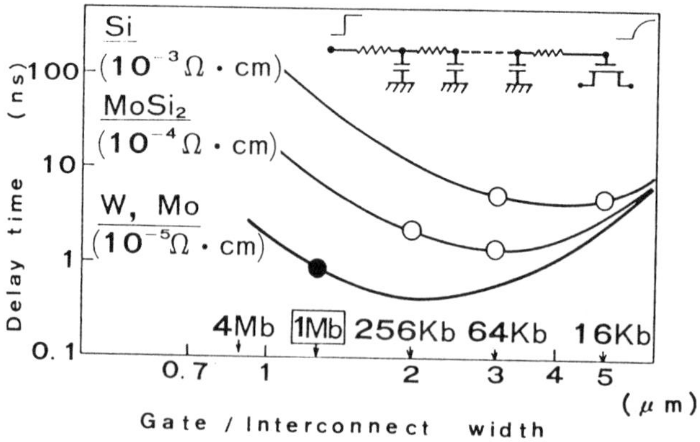

Figure 1: Future generation MOS VLSI gate electrode and interconnect material choices.[4] Reprinted by permission of the publisher, The Electrochemical Society, Inc.

4.2 REFRACTORY METAL SILICIDES

In this section, we will restrict our attention to the silicides of tungsten, molybdenum, tantalum and titanium. CVD WSi_2 is currently being used commercially, and the other three have either been considered seriously or used to a limited extent. We will start with WSi_2.

4.2.1 Tungsten Silicide

Using a cold-wall CVD reactor similar to the internally-heated barrel described in Figure 22 of Chapter 2, tungsten silicide was deposited from WF_6 and SiH_4,[5] which is often described by the overall reaction

(1) $$WF_6 + 2SiH_4 \rightarrow WSi_2(s) + 6HF + H_2.$$

Experimental evidence shows, however, that there is very little HF found as a byproduct when $WSi_2(s)$ is deposited from WF_6 and SiH_4.[6] From thermodynamic considerations, we could anticipate reaction products such as SiF_4, $SiHF_3$, SiH_2F_2, SiH_3F, SiF_2, HF and H_2 as well as possibly others. Therefore, a more accurate representation would be

$$(WF_6, SiH_4) \rightarrow (WSi_2(s), HF, SiF_4, SiHF_3, SiH_2F_2, SiH_3F, SiF_2, H_2).$$

Depositions were done over a temperature range of 330° to 450°C and a pressure range of 50 to 300 mTorr. A SiH_4 to WF_6 ratio of 70:1 was used, which resulted in a Si:W ratio of 2.2 to 2.7. In other words, they achieved films with the composition of WSi_x where $2.2 < x < 2.7$. Deposition rates varied with WF_6 flow rate, as shown in Figure 2. On the other hand, they did not vary with pressure or deposition temperature. The stoichiometry of the as-deposited film also varied with the WF_6 flow rate, as shown in Figure 3.

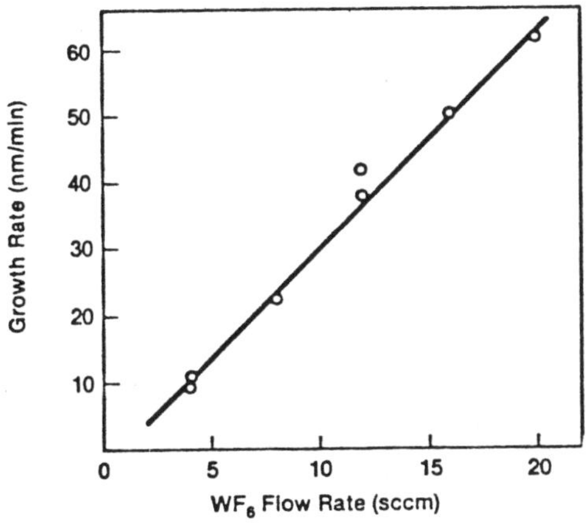

Figure 2: Deposition rate of WSi_x versus WF_6 flow rate.[7]

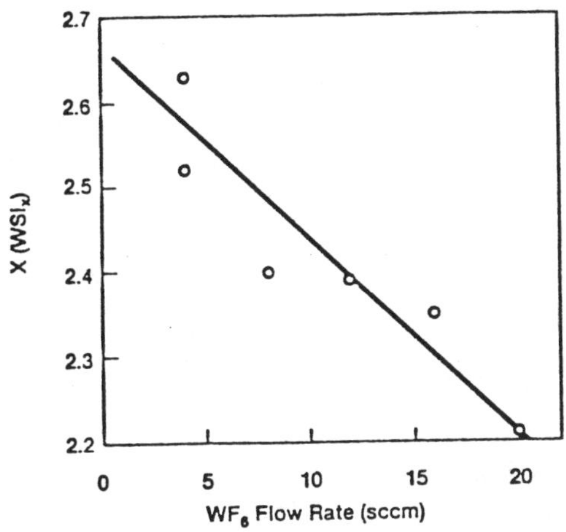

Figure 3: Film stoichiometry as a function of WF_6 flow rate.[7]

As mentioned earlier, the "polycide" structure can be used to replace the traditional gate poly. A sketch of such a configuration is shown in Figure 4, when $x > 2.0$, the WSi_x is stable on the poly. Otherwise, it cracks and/or peels off during high temperature processing.

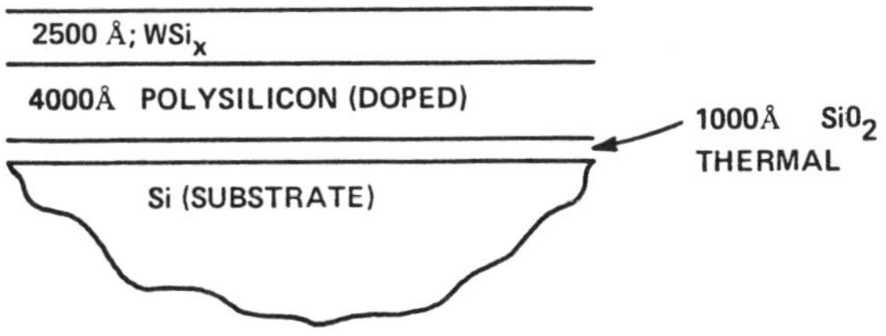

Figure 4: Polycide structure.

The as-deposited WSi_x exhibits high resistivity, which drops to acceptable values upon heat treatment. The explanation is that the as-deposited film has very small grains and they grow to quite large grains as a result of the heat treatment. This phenomena is illustrated in Figure 5, where 1000 Å thick films were deposited on bare (100) silicon wafers and annealed for 1 hour in argon at different temperatures. The as-deposited film has a microcrystalline structure with 30 Å grains. When annealed at 1000°C, the film became polycrystalline, with crystals 750 Å across.

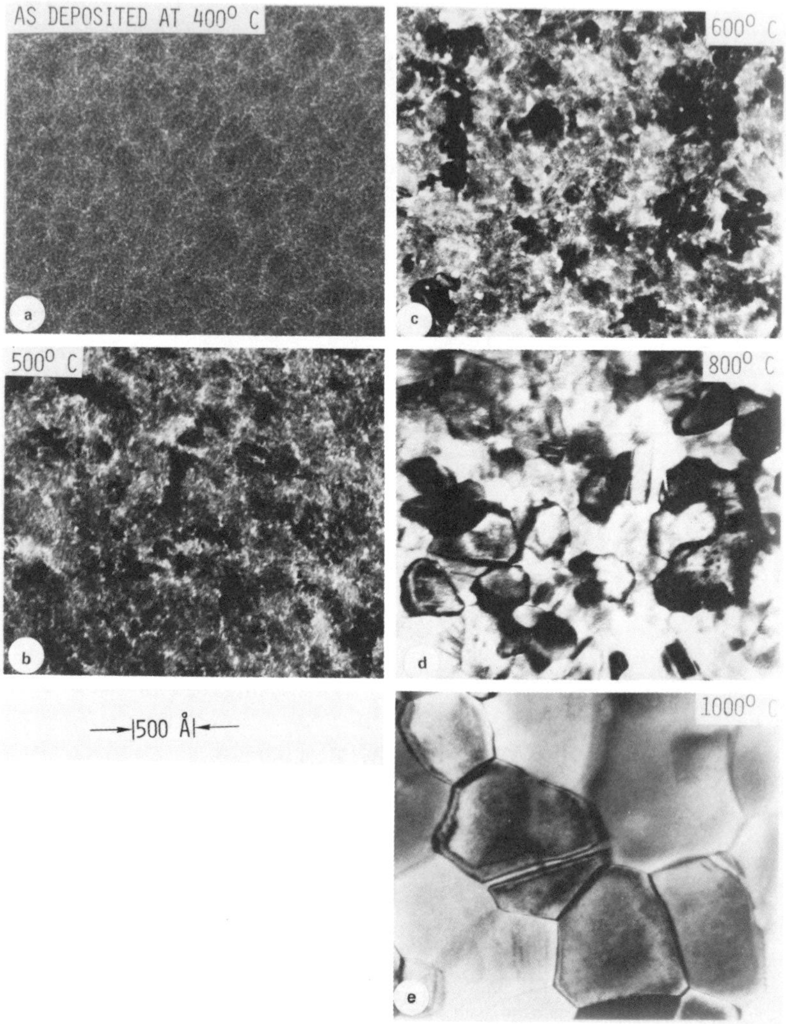

Figure 5: TEMs of 1000 Å LPCVD WSi_2:[5] (a) as-deposited, 400°C, or annealed, (b) 500°C, (c) 600°C, (d) 800°C, and (e) 1000°C.

The same effect is seen when the polycide structure of Figure 4 is annealed. In Figure 6 we see the effect of furnace annealing; Figure 7 shows similar effects for rapid thermal annealing.

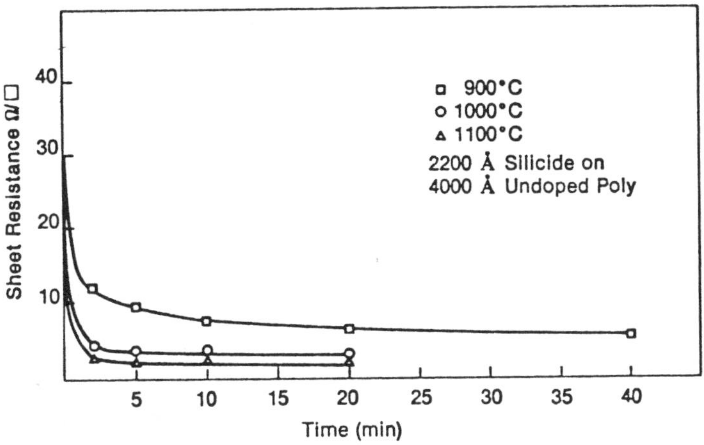

Figure 6: Resistivity of WSi_x films on polysilicon with furnace anneals.[7]

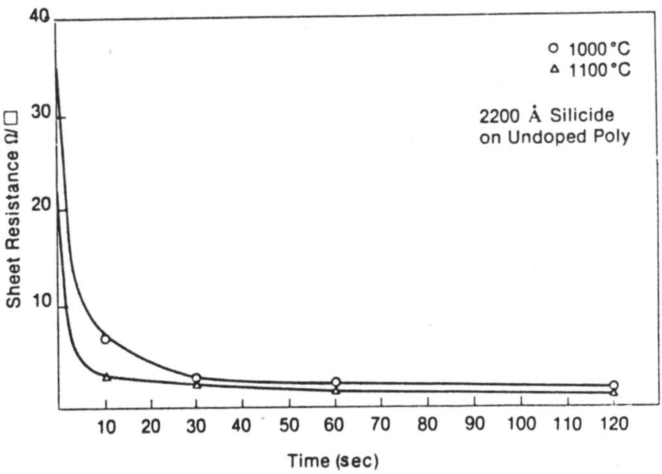

Figure 7: Resistivity of WSi_x films with rapid anneals.[8]

Another feature of the annealed films is that their final resistivity is dependent on their unannealed stoichiometry. This effect is illustrated in Figure 8, where we see that the lowest resistivity film is obtained with a slightly silicon-rich film ($WSi_{2.2}$). Apparently during the annealing process, some of the Si in the $WSi_{2.2}$ migrates down to the poly underlayer, leaving this poly slightly thicker (about 150 Å) and the silicide closer to WSi_2, as desired.

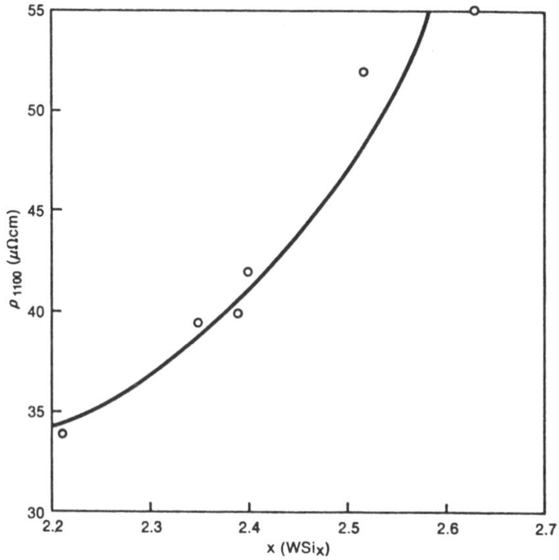

Figure 8: Annealed resistivity versus WSi_x stoichiometry.[7]

The tungsten silicide films, when deposited by CVD as contrasted to sputtering, are very conformal. This behavior is shown in Figure 9, where a 5000 Å polycide film (2500 Å WSi_2 + 2500 Å poly) is deposited over a small step. On steps with vertical walls, the thickness of the polycide on the vertical wall was at least 75% of its thickness on the horizontal surface.

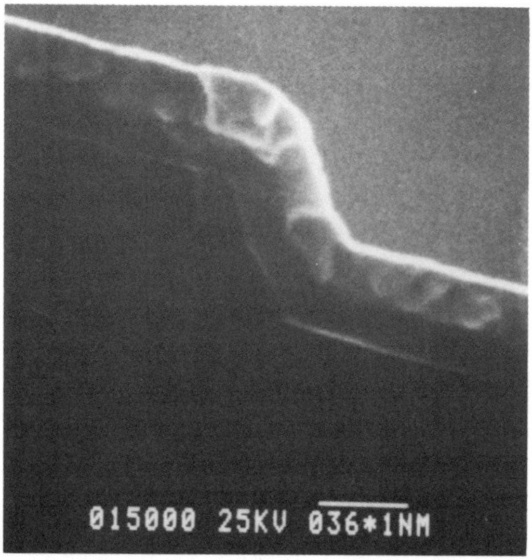

Figure 9: SEM of step coverage of CVD WSi_2.[5]

One of the primary advantages of the polycide concept is that the silicide top layer oxidizes readily to form a dense adherent SiO_2 overlayer, and the polycide structure underneath remains intact. Also, the oxide film forms within a reasonable time. Oxide thicknesses formed by dry O_2 oxidation are shown in Figure 10, as a function of time and temperature.

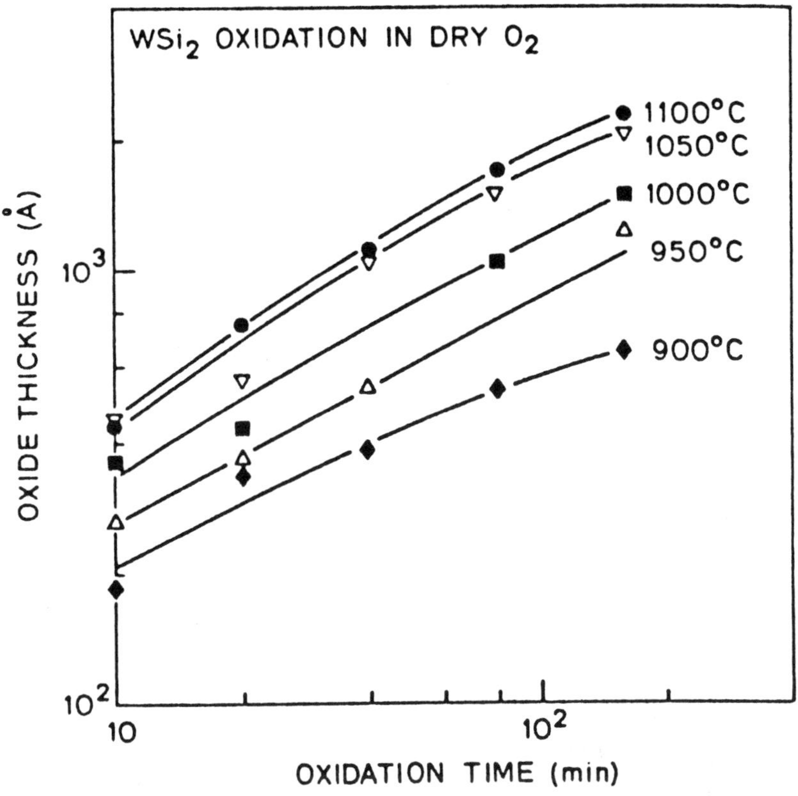

Figure 10: Oxide thicknesses for dry oxidation of WSi_2.[5]

The mechanism whereby the silicide acquires its SiO_2 overlayer is important to the understanding of the value of this structure. Basically, Si diffuses from the poly layer up to the surface of the WSi_2 layer where oxidation begins. As the oxide grows, O_2 diffuses from the surface of the oxide down to the interface between the oxide and WSi_2. At the same time, Si continues to diffuse up to this same interface from the poly layer. In fact, if a very thick oxide layer is grown, and the underlying poly were fairly thin, it is possible to completely consume the polysilicon layer. Obviously, it would neither be necessary or desirable to carry the process this far. The fact that the WSi_2 layer remains intact during this process allows the integrated circuit designer to continue to use the usual process steps while obtaining the benefits of the lower resistivity polycide layer.

4.2.2 Molybdenum Silicide

Compared to the detailed studies of WSi_2, relatively less has been reported on LPCVD of the other refractory metals. One study in which $MoSi_2$ has been deposited from $MoCl_5$, SiH_4 and H_2 has been described recently.[9] In this work, a hot tube furnace was used with a $MoCl_5$ sublimator using an H_2 carrier gas attached to it. The sublimator was operated at 160°C and all lines leading to the furnace were heat traced to prevent $MoCl_5$ condensation.

A typical deposition was carried out with $MoCl_5$ in H_2 gas at 190 sccm, SiH_4 at 210 sccm, and N_2 as a diluent gas. Pressures were in the range of 600 mTorr to 2.0 Torr. At a temperature of 670°C, a deposition rate of 180 Å/min was achieved. Although several molybdenum silicide phases are known (Mo_3Si, Mo_5Si_3, and $MoSi_2$), only $MoSi_2$ was found in these experiments.

The as-deposited films had resistivities of 1000 $\mu\Omega$-cm. After annealing at 1000°C for 20 minutes, this reduced to 120 $\mu\Omega$-cm, which compared to the WSi_2 described earlier is more than twice the value. One advantage of this film is that some small amount of chlorine remains after deposition, and this chlorine can act as a getter for any mobile impurities such as Na.

Another recent study has examined the $MoSi_2$ deposition using MoF_6 and SiH_4 rather than the chloride.[10] These experiments were carried out in a quartz cold-wall system. Best results were found at a deposition temperature of 150°C, which is much lower than needed by the chloride process. Typical deposition conditions were 250 mTorr pressure, 2 to 6 sccm of MoF_6 and 100 sccm of SiH_4. All films were annealed at 1100°C for 30 seconds. They were found to be very stable and adherent on both silicon and silicon dioxide. A minimum resistivity of ~100 $\mu\Omega$-cm was found for deposition temperatures below 150°C. Compared to the chloride process, the major advantage is the low temperature at which these films can be deposited.

Finally, the conformality of the $MoSi_2$ films are good.

4.2.3 Tantalum Silicide

The WSi_2 films described earlier were deposited in a cold-wall reactor. The $MoSi_2$ films just described were deposited in a hot-wall reactor. Does it make any difference whether these films are deposited in one or the other type of reactor? In studying the deposition of $TaSi_2$, we have an opportunity to examine this question, as we have two studies to consider; one was done in a hot tube[11] and the other was done in a cold-wall system.[12,13] Both study the reaction between $TaCl_5$ and SiH_4 at comparable temperatures and pressures. The desired tantalum silicide is $TaSi_2$, which has a low electrical resistivity and is thermodynamically stable.

The hot-wall study[11] introduced $TaCl_5$ with an evaporator operating in the temperature range of 120° to 140°C with a small H_2 flow (about 5 sccm) as the carrier gas. A SiH_4 flow of 24 sccm is used at a pressure of 280 mTorr. Deposition rates of 120 Å/min were achieved at temperatures of 615° to 635°C, with uniformity of ± 10%. Actually, a polysilicon layer is deposited first, so that a polycide structure could be studied.

Unfortunately, the tantalum silicide deposited was very metal rich, and close to Ta_5Si_3. Although attempts to vary the stoichiometry by changing the process conditions were made, they were unable to report the deposition of $TaSi_2$ films.

Instead, the approach taken was to anneal the polycide films at temperatures over 800°C. In this case, Si diffused up to the Ta_5Si_3 film in sufficient quantity to convert it to $TaSi_2$. An anneal at 1000°C for 15 minutes in argon produced a resistivity of 48 $\mu\Omega$-cm in 2500 Å thick $TaSi_2$ layers.

Although these results were encouraging, examination of the films by TEM showed that almost all of the underlying poly was consumed in creating $TaSi_2$. Also the Si was not extracted uniformly from the poly layer, so that the surface of the $TaSi_2$ was quite rough.

The second study was done in a cold-wall reactor[12,13] using the same reactants. The reactor was a single-wafer system, similar to the tube reactor of Figure 18 in Chapter 2, with the wafer heated by an electrical resistance heater in the pedestal. In this case, the sublimator was operated at 88°C with a 10 sccm flow of H_2. The influence of SiH_4 flow rate on the film stoichiometry and resistivity (after anneal) are shown in Figure 11.

Films of $TaSi_2$ deposited in this process, after anneal, were specular (surface roughness of only 1000 Å grain structure) and had good step coverage (thickness on vertical wall equals 65% of thickness on horizontal surface).

As long as the films were not Si rich, the resistivity was in the range of ~75 $\mu\Omega$-cm. When substrate temperature was varied between 650°C and 750°C, the deposition rates were unchanged. This implies that the reaction is proceeding by a diffusion-controlled mechanism. The resistivity of the better films after a 1-hour, 900°C anneal in argon was ~60 $\mu\Omega$-cm, independent of the deposition temperature.

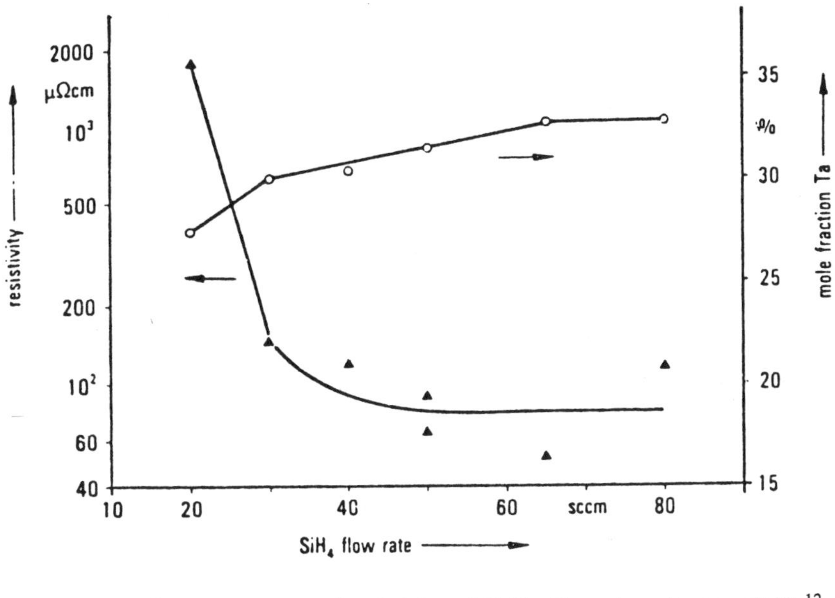

Figure 11: Stoichiometry and resistivity of $TaSi_2$ films deposited by CVD.[12]

An interesting effect was observed by varying the pressure between 80 and 400 mTorr. At pressures of 180 mTorr and above, the deposition rate jumped from 600 to 2000 Å/min. At the same time, resistivity rose as high as 4000 $\mu\Omega$-cm. The variation with pressure is shown in Figure 12. Apparently, the stoichiometry changed dramatically at pressures of 180 mTorr and higher. In fact, there is very little Ta in the film created at 400 mTorr (about 13%), so this is mostly a polysilicon film.

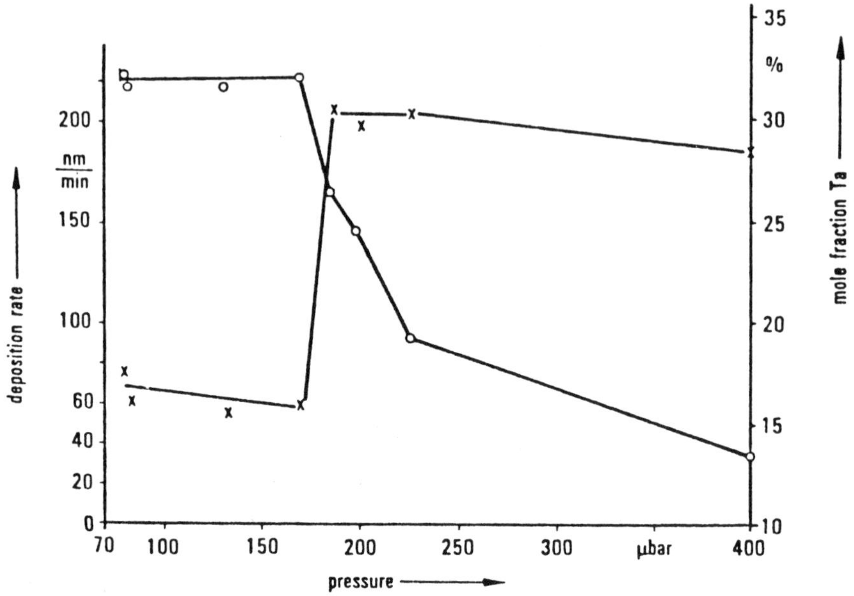

Figure 12: Resistivity and stoichiometry of $TaSi_x$ CVD films as a function of pressure.[12]

Finally, we can comment on the influence of the reactor type on the films that can be deposited. Evidently, the hot-wall reactor tends to deposit very Ta-rich films. Although it may be possible to alter the stoichiometry in this type of reactor, the choices are limited. One must operate under conditions where uniform depositions are achieved both on each wafer and from wafer to wafer, because this is a batch system. In the cold-wall reactor, it was possible to obtain the proper stoichiometry at high deposition rates. Since the higher deposition rates permit development of a single-wafer reactor, there are more choices in the process conditions to be used.

It is probable that a fundamental difference exists in processes operating in the two types of reactors considered here. In the hot-wall system, the reactant gases have ample time to react before reaching the wafers, so gas phase chemistry probably plays a role. In the cold-wall system, this is probably minimized.

4.2.4 Titanium Silicide

The lowest resistivity silicide film of the four we are considering is the $TiSi_2$ film, so such films have always been of interest. A recent study[14] has shown that these films can also be deposited by low-pressure CVD. For these experiments, a cold-wall reactor similar to the parallel-flow tube reactor sketched in Figure 17 of Chapter 1 was used. The wafer was heated by heating the susceptor from below by optical radiation.

Depositions were carried out with SiH_4 and $TiCl_4$ reactants. The $TiCl_4$, which is a liquid at room temperature, is evaporated in a sublimator at 28°C. The structure grown was the polycide structure, as before. Films were deposited at 650° to 750°C and pressures from 50 to 460 mTorr at several flow rates ($TiCl_4/SiH_4$). Stoichiometric films that were slightly Si rich ($TiSi_2$) were achieved with as-deposited resistivity of 22 $\mu\Omega$-cm reported. Also, surface roughness was small (about 50 Å).

In summary then, good quality $TiSi_2$ films were produced with low as-deposited resistivities. The only concern, as far as using such films is concerned, is the fact that $TiSi_2$ etches readily during wet HF etch procedures. Such etch procedures are an integral part of many of the integrated circuit process steps, and one must be concerned about the integrity of the $TiSi_2$ films. If the IC manufacturer is willing to use all dry etch procedures (plasma etching), this concern can be alleviated.

4.3 TUNGSTEN

As mentioned earlier, there is a considerable need for a conformal metallic coating with resistivity close to that of aluminum, but with a higher melting point. Of the ones we have been considering, the two lowest resistivity candidates are molybdenum and tungsten. Tungsten has received the most attention since the H_2 reduction of WF_6 process has been under development for a variety of applications since 1967.[15] The extent of current interest can be seen in a recent publication.[16]

There are two aspects of tungsten CVD for integrated circuits that have taken on commercial importance. One is the blanket deposition and subsequent patterning, so it can be used as a conductor to replace high-resistivity doped poly. The second area of interest is the "selective" CVD of tungsten, where deposition occurs on silicon but not on silicon dioxide. Here one can selectively fill via holes to either provide a thin barrier metal or to deposit a thicker layer to help planarize the circuit. Both applications involve only one processing step, and are attractive for this reason.

We will review recent work in the blanket tungsten process first.

4.3.1 Blanket Tungsten

Tungsten can be deposited by CVD by a number of different processes. Several that have received considerable study are:

$$(2) \qquad WF_6 + 3H_2 \rightleftharpoons W(s) + 6HF$$

(3) $2WF_6 + 3Si(s) \rightleftharpoons 2W(s) + 3SiF_4$

(4) $WCl_6 + 3H_2 \rightleftharpoons W(s) + 6HCl$

(5) $2WF_6 + 3SiH_4 \rightleftharpoons 2W(s) + 3SiF_4 + 6H_2$

The first is the hydrogen reduction process which can proceed on any surface raised to a suitable temperature. The second is the silicon reduction process where silicon reduces WF_6. The third process is similar to the first, but substitutes chlorine for fluorine. The final process is related to the WSi_2 deposition studied earlier. It has been shown[17] that depending on the deposition conditions, one can deposit either W, WSi_2 or W_5Si_3 from these two reactants.

The second reaction, Equation (3), is the basis for the selective tungsten process we will discuss later. It also plays some role in the blanket process.

The first study of CVD tungsten for application to integrated circuits was done by Shaw and Amick,[18] working with the hexafluoride. They carried out their depositions in an atmospheric-pressure horizontal cold-wall tube reactor (see Figure 17, Chapter 1), where the susceptor that held the wafers was inductively heated.

The major problem, then and now, in attempting to deposit blanket tungsten is the adhesion to silicon dioxide. Unless some steps are taken beforehand, the H_2 reduction of WF_6 on SiO_2 will not produce an adherent film. The solution to this problem was to undertake the selective process initially.[18] It is then pointed out that WF_6 will decompose on Si, but it will slowly etch SiO_2. Once the SiO_2 has been etched by WF_6, then the W deposited from the H_2 reduction process adheres very well.

The two-step process just described was operated at 700°C and atmospheric pressure. Films up to 0.5 micron were reported to be specular with no indication of cracking or loss of adhesion to the oxide. A resistivity of 6 μΩ-cm was achieved.

More recently, cold-wall blanket W depositions have been done at low pressures.[19,20] In this case, the problem of poor adhesion to the oxide surface was solved by depositing an intermediate layer of WSi_2. The silicide adheres very well to oxide, and if properly treated, the W will adhere to the silicide. One suggested treatment is a plasma etch of the freshly deposited WSi_2 with NF_3. All depositions were carried out at 500° to 600°C and pressures of 200 to 500 mTorr in the same barrel reactor used for tungsten silicide studies.[8] The silicide layer was typically 10% of the thickness of the final layer. Accordingly, the silicon content was approximately 5% and the film resistivity was 8 to 10 μΩ-cm for 1-micron thick films. Good conformality of these films was obtained, as can be seen in Figure 13.[19]

The hydrogen reduction of the hexafluoride process has also been developed in a hot wall process.[21] Experiments were carried out in a traditional diffusion furnace tube with deposition rates approximately 100 Å/min. Resistivities of 14 μΩ-cm were found for 2000 Å thick pure tungsten films. Adhesion of these films on silicon was reported to be good, but not good on silicon dioxide.

Tungsten hexachloride has not been studied as extensively as the fluoride for tungsten deposition, since the chloride is a solid at room temperature and must be heated to 170°C to achieve a reasonable vapor pressure. Of course, all

Figure 13: CVD tungsten on silicon.[19]

lines leading from the sublimator to the reactor must also be heated to 170°C to prevent condensation. In one experiment,[22] WCl_6 was reduced by H_2 in an atmospheric pressure reactor similar to the horizontal cold-wall reactor used by Shaw and Amick.[18] Depositions were carried out at 600°C and deposition rates of 100 Å/min were achieved. No information on the quality of the tungsten film was reported, however.

Finally, blanket tungsten can be deposited from tungsten carbonyl, $W(CO)_6$. Some data has been published and such depositions appear to result in films with appreciable carbon and oxygen.[23]

4.3.2 Selective Tungsten

Although there are a number of reactions from which the selective deposition of tungsten could be implemented, the only ones that have been investigated are the reduction of WF_6 by either Si or H_2 [see Equations (5) and (6)]. The first of these will obviously be selective. The second can be selective when the H_2 reduction is favored on W surfaces, as compared to oxide surfaces.

The most comprehensive experiments have been performed in low-pressure CVD hot tube reactors.[24] When WF_6 is reduced on clean, flat silicon surfaces, the deposition rate is very rapid (>1000 Å/min) and self limiting. Generally, a tungsten film of less than 200 Å (grown at 300° to 425°C and 500 mTorr) is sufficient to completely block this reaction, as shown in Figure 14.

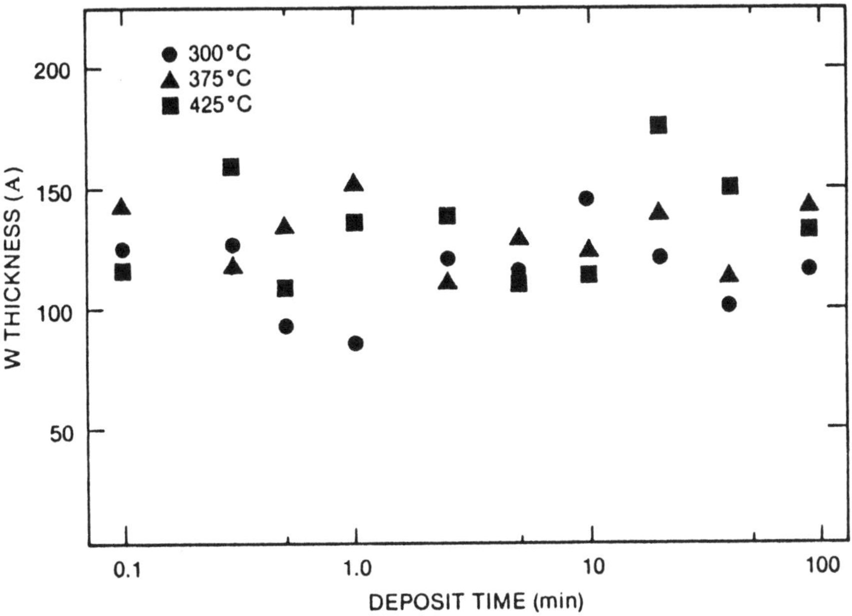

Figure 14: Silicon reduction of WF_6–thickness of tungsten versus deposition time.[24] Reprinted by permission of the publisher, The Electrochemical Society, Inc.

If a selective tungsten layer thicker than 200 Å is required, selectively deposited, then the two-step process originally suggested by Shaw[18] can be used. In this procedure, shortly after the limiting tungsten layer has been deposited, H_2 is added to the reactant gases and tungsten deposition continues. This scheme is effective as long as there is no simultaneous reduction on adjacent dielectric surfaces. Alternately, the H_2 can be added to the reactant gases at the start, and the same process will occur[24] since Si is a much more effective reducing agent for WF_6 than H_2. This phenomenon is shown clearly in Figure 15.

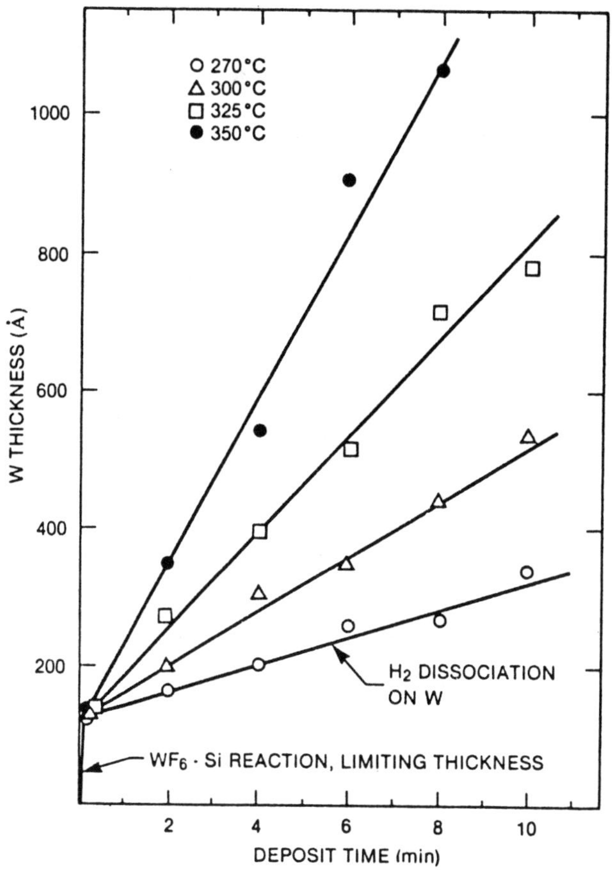

Figure 15: Tungsten deposit thickness as a function of time for several temperatures.[24] Reprinted by permission of the publisher, The Electrochemical Society, Inc.

Since the mechanism whereby silicon reduces WF_6 involves the consumption of silicon (i.e., production of SiF_4, SiF_2, etc.), there is concern that there should not be too much encroachment of the tungsten into the silicon. Assuming the silicon is consumed uniformly, then two atoms of silicon will be released for each tungsten atom deposited. As tungsten is a denser material than Si, this translates into a thickness of Si consumed which is twice the thickness of tungsten deposited.

The Si consumption can be beneficial in that a more intimate contact between W and Si is promoted.[25] It can, however, lead to encroachment problems that will be discussed later. As in the blanket tungsten deposition case, film resistivities are higher than bulk. For selective deposition, thinner films are useful, so it is important to recognize that resistivity varies with film thickness, as shown in Figure 16.[26]

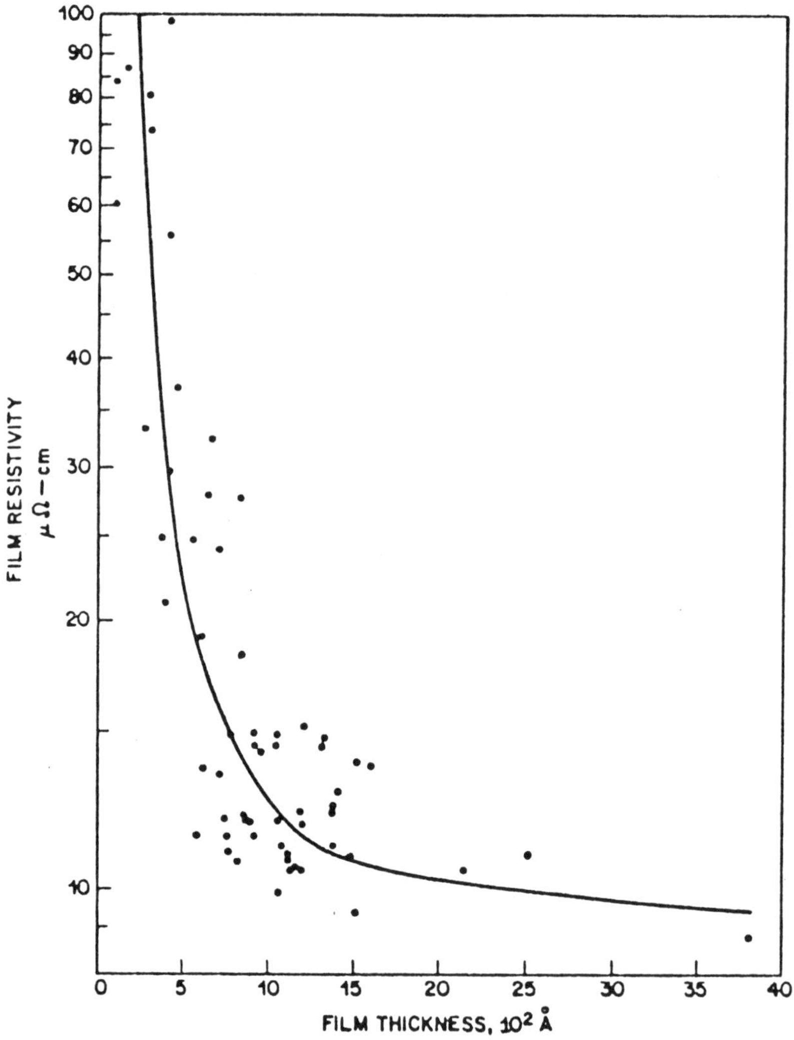

Figure 16: Tungsten film resistivity versus film thickness.[26] Reprinted by permission of the publisher, The Electrochemical Society, Inc.

It is important to recognize that substrate pretreatment is important for selective tungsten depositions. The character of the silicon surface, and its oxide coating, play a decisive role in tungsten growth behavior. When starting with a clean Si substrate and only a thin native oxide or no oxide on it, silicon reduction yields good quality films that self limit at approximately 200 Å. If the substrate is treated with a glow discharge (CF_4/O_2), a rough 200 Å oxide layer is formed. In this case, the limiting tungsten layer can be thicker, but it is not a good quality surface.[27] This phenomenon is illustrated in Figure 17.

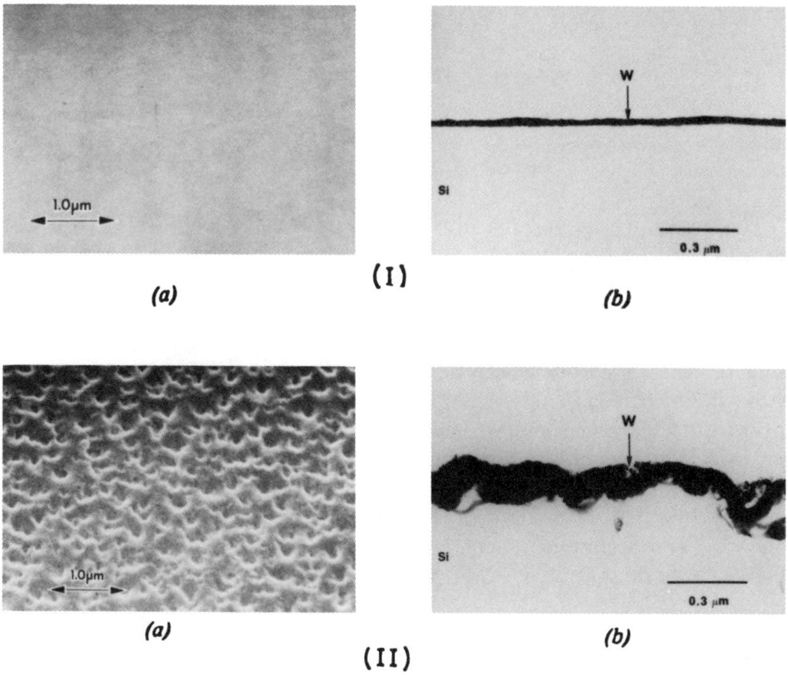

Figure 17: Photographs of two limiting tungsten films: (I) Tungsten deposited with dilute HF preclean; (II) Tungsten deposited with CF_4/O_2 glow discharge clean; (a) SEM; (b) TEM.[27]

Depending on deposition conditions, tungsten may nucleate and deposit on dielectric films. In order for a selective process to be successful, such deposition must be prevented. Qualitatively, higher deposition temperatures and higher concentrations of WF_6 in the reactant flow promote nucleation and deposition. However, even if low temperatures ($<500°C$) and a dilute reactant gas stream ($H_2/WF_6 = 220$) are used, selectivity can be lost under certain conditions. It has been observed[27] that where there are large areas of exposed silicon on a wafer, that selectivity is lost on oxide adjacent to these areas. Therefore, it has been suggested that reaction by products from the selective deposition of tungsten (SiF_4, HF) may cause this loss of selectivity. Backside coating of wafers with oxide, for example, seems to be helpful in maintaining selectivity. For depositions up to about 3000 Å, it appears that good selectivity is achieved. To date, experimental results have been inconsistent for thicker films.

Finally, the type of dielectric also plays a role in selectivity. The percentage monolayer coverage achieved in a WF_6 deposition at 300°C for 40 minutes is shown in Table 2 for several different films.[24]

Table 2: Monolayer Coverage of Selective Tungsten on Different Dielectrics[27]

PECVD Si_3N_4	12%
LPCVD Si_3N_4	5.2
Thermal SiO_2 (p-doped)	1.3
Atmos. PSG (7% P)	0.86
Thermal SiO_2 (undoped)	0.40

The remaining important phenomena that occur with selective tungsten deposition are "encroachment" and "tunneling."[28] When the silicon in a contact hole reduces WF_6, it is covered by a layer of tungsten. When the layer becomes thick enough so that WF_6 can no longer diffuse through the tungsten layer, the reaction stops. However, consider an idealized contact hole as shown in Figure 18. The corner where Si, W and SiO_2 meet is illustrated in expanded fashion in (b) of Figure 17. Observe that the reaction proceeds by WF_6 diffusing down through the tungsten, reacting at the silicon surface and then SiF_4 diffuses back out through the tungsten. When the tungsten film becomes too thick, an alternate path for WF_6 to find silicon is at the corner junction. Accordingly, when the WF_6 flow is maintained long after the limiting thickness of W has been achieved, "encroachment" occurs. Such behavior is illustrated in Figure 19 for a WF_6 + Ar deposition at 300°C for 40 minutes.

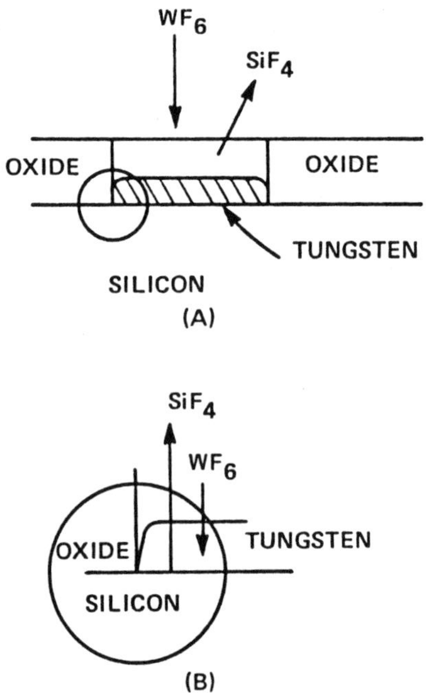

Figure 18: Selective tungsten deposition in a contact hole.

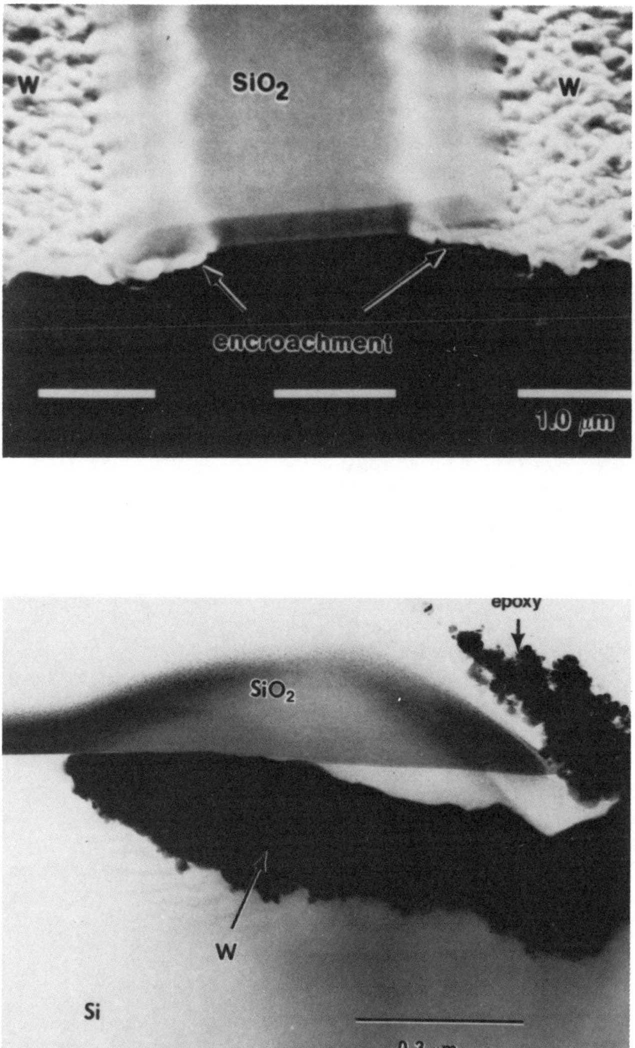

Figure 19: Encroachment of tungsten under SiO_2.[28] Top figure is SEM; bottom one is TEM. Reprinted by permission of the publisher, The Electrochemical Society, Inc.

A phenomenon related to encroachment is the appearance of "tunnels." One example of such tunnels is shown in Figure 20. There appear to be many individual tunnels, each of constant diameter (200 to 400 Å), with a single tungsten-containing particle at the end. The tunnels are observed when the same process is done on aluminum. They also occur whether or not H_2 is added to the WF_6. Therefore, it appears that SiF_4 or SiF_2 (which are gaseous) must be the product of whatever reaction is causing the tunnel formation.

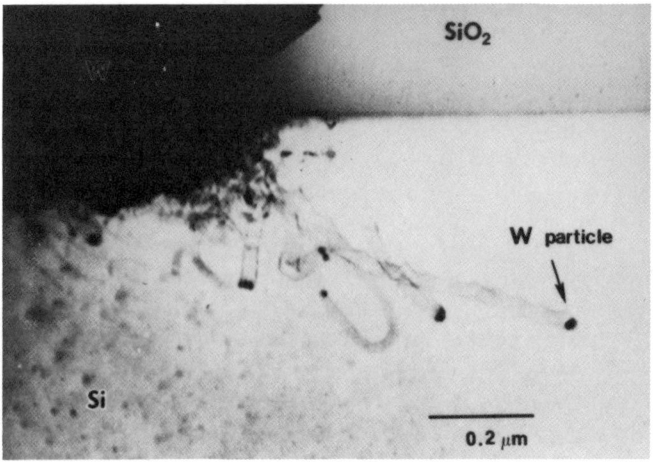

Figure 20: TEM illustration of tunnel formation for selective deposition at 300°C.[28] Reprinted by permission of the publisher, The Electrochemical Society, Inc.

When the selective deposition is carried out at different temperatures, the degree of encroachment varies.[29] This is illustrated in Figure 21 for a short 3-minute deposition.

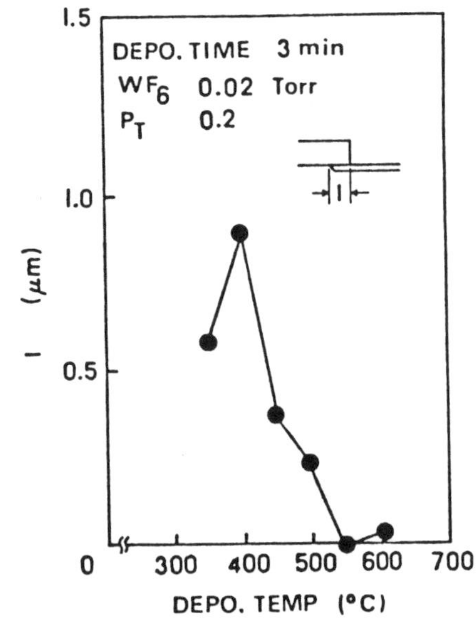

Figure 21: Encroachment length at different deposition temperatures.[29]

It appears that if the deposition temperature is chosen precisely at 550°C, that there is no encroachment. In fact, if the deposition is continued beyond the time when the limiting W thickness has been achieved, then "creep up" occurs. This is shown in Figure 22.[29] Additional W is deposited, but it occurs along the outer surface of the adjacent oxide, rather than at the silicon/silicon dioxide interface.

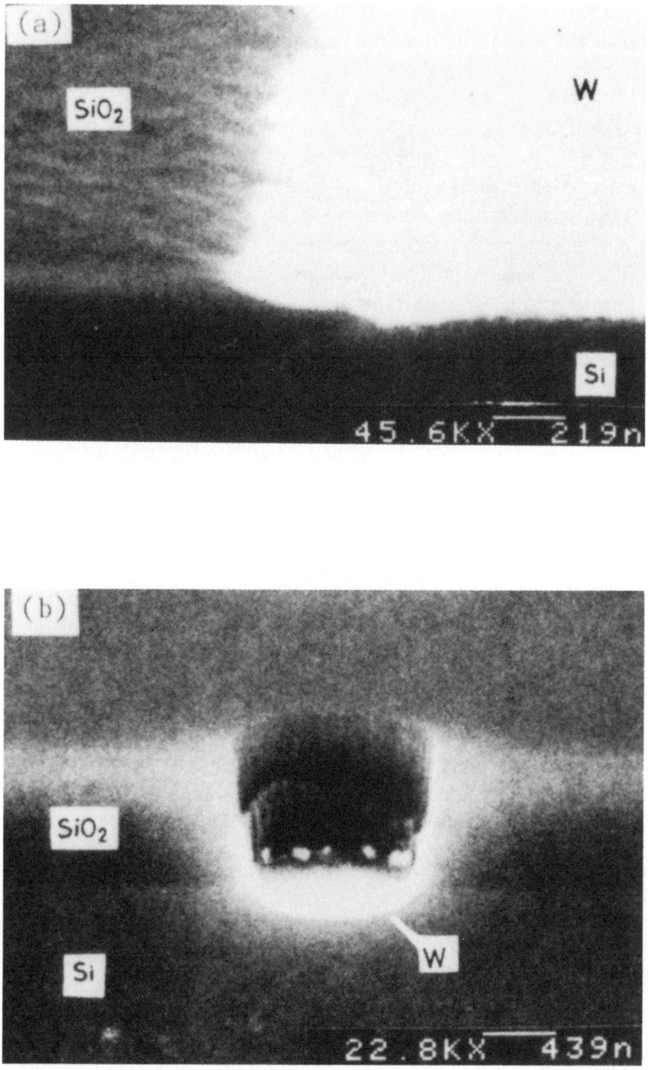

Figure 22: Creep-up phenomena onto (a) field oxide and (b) the wall of contact hole.[29]

In contrast to the many studies of selective tungsten deposition in a hot wall reactor done to date, only one has been done in a cold-wall reactor.[30] It was done in a laboratory-scale, horizontal tube with wafer heating done by radiation from high-intensity tungsten filament lamps. Selective depositions done at temperatures as high as 520°C (p ~300 to 800 mTorr) were done at rates in excess of 1000 Å/min. Such high rates would be particularly valuable for via hole filling, where on the order of 1 micron of material has to be deposited. It should be noted that although deposition on Mo, Al and Si are mentioned, the test data reported were for depositions on Mo. These authors also reported a more rapid growth of W up the side of their contact holes, which may be related to the creep-up phenomena mentioned earlier.[29]

4.4 ALUMINUM

Evaporated and sputtered aluminum films have been successfully used in integrated circuit manufacture for some years. Problems due to spiking of aluminum into the silicon it is contacting have been minimized by doping the films with 0.5 to 1.0% Si. Again, aluminum is susceptible to electromigration failure due to aluminum transport at high current densities. That is, at high current densities through a narrow conductor, the conductor material can be transported in the direction of current flow. In the extreme, this results in opens occurring in these conductors. Such problems have also been minimized by introducing 1 to 4% Cu into the films. The primary shortcoming of these aluminum films is their lack of conformality when attempting to cover 1 micron steps or trenches. This is due to the inherent line of sight nature of the evaporation and sputtering techniques.

Since CVD processes offer the possibility of conformal coverage of a severe terrain, this method of depositing aluminum has been considered. The most completely explored method employs an aluminum-organic compound, TIBAL,[31,32] which is

$$Al(C_4H_9)_3 \rightarrow \text{Tri-isobutyl-aluminum}$$

and the overall reaction is thought to be

$$\text{(6)} \qquad Al(C_4H_9) \rightarrow Al + \tfrac{3}{2}H_2 + 3C_4H_8.$$

Depositions were carried out in a hot-wall tube reactor. Since TIBAL is a liquid with a vapor pressure of 1.5 Torr at 45°C, it was introduced into the reactor by heating in an evaporator. Special precautions were taken because TIBAL is pyrophoric.

Prior to the aluminum deposition, it was found necessary to "activate" the wafer surface by exposing it to $TiCl_4$ vapor. Presumably, Ti was deposited, and the authors claim that this served to provide numerous nucleation sites for the pyrolysis of the TIBAL. Deposition was carried out at temperatures in the range of 220° to 300°C and pressures were, typically, 200 to 500 mTorr. Film growth rates were about 200 to 800 Å/min. Incorporation of Si into the films was achieved by exposing the wafers to $SiH_4 + H_2$ (10 Torr) in the same reactor. Apparently, no attempt was made to add Cu to the films.

The films were adherent, and no appreciable quantities of C, H or other impurities were detected. The resistivity was measured to be only 10% higher than bulk aluminum. As shown in Figure 23, the films appeared to be quite conformal.

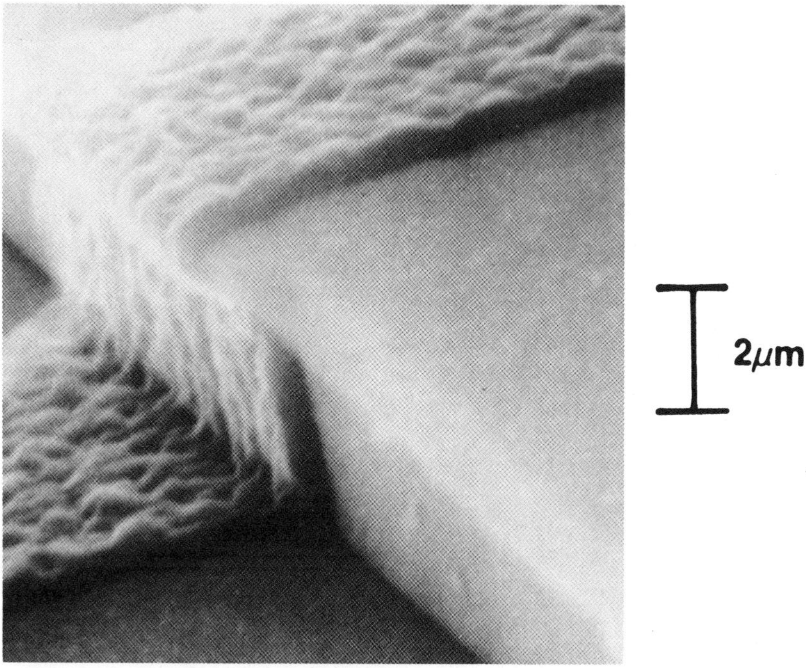

Figure 23: CVD aluminum over vertical oxide step with overhang.[31]

The primary difference between these CVD films and those obtained by evaporation or sputtering was their roughness. The films produced were not specular. They had a peak-to-peak surface roughness of 1000 to 1500 Å. SEM pictures of a film deposited at 270°C and 0.4 Torr (see Figure 24) show this roughness.

In summary, CVD aluminum films produced by low-pressure pyrolysis of TIBAL have been shown to achieve the improved conformality desired. Unfortunately, they are rougher than standard films, and no feasible way of introducing Cu into the film has been found to prevent electromigration.

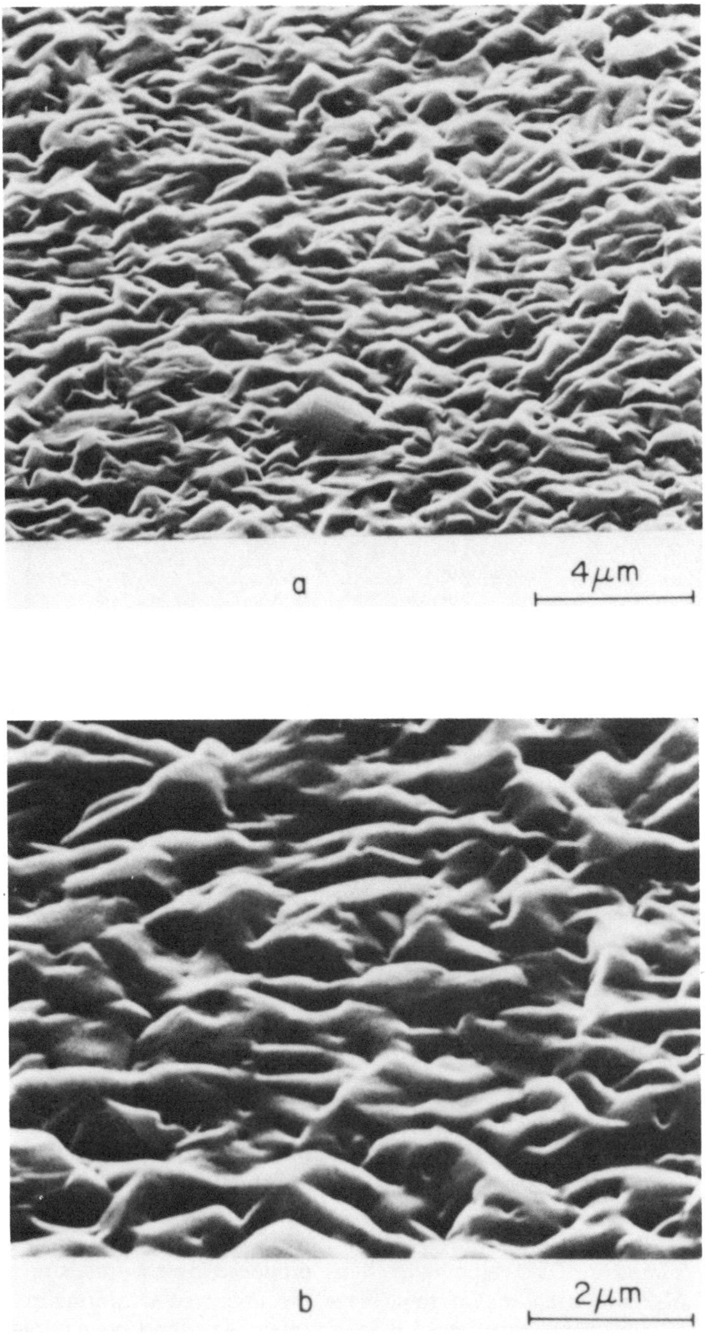

Figure 24: SEM photos of CVD aluminum film.[32]

REFERENCES

1. Kern, W., Chemical Methods of Film Deposition, in *Thin Film Processes,* eds. J.L. Vossen and W. Kern, Academic Press, NY (1978)
2. Saraswat, K.C. and Mohammadi, F., Effect of interconnection scaling on time delay of VLSI circuits, IEEE Trans. Electron Devices. ED-29: 645 (1982).
3. Murarka, S.P., *Silicides for VLSI Applications.* Academic Press, NY (1983).
4. Yamamoto, N., Kume, H., Iwata, S., Yagi, K., Kobayashi, N., Mori, N. and Miyasaki, H., Fabrication of highly reliable tungsten gate MOS VLSI's. *J. Electrochem. Soc.* 133:401 (1986).
5. Saraswat, K.C., Brors, D.L., Fair, J.A., Monnig, K.A. and Beyers, R., Properties of low-pressure CVD tungsten silicide for MOS VLSI interconnections. *IEEE Trans. on Electron Dev.* ED-30:1497 (1983).
6. Private communication, Gaczi, P., and Reynolds, G.
7. Brors, D.L., Fair, J.A., Monnig, K.A. and Saraswat, K.C., Deposition parameters and properties of low pressure chemical vapor deposited tungsten silicide for integrated circuits manufacture, in *Proceedings of the Ninth International Conference on Chemical Vapor Deposition.* eds. Robinson, McD., van den Brekel, C.H.J. Electrochem. Soc., Pennington, NJ (1984), p. 283. These figures were originally presented at the Spring 1984 Meeting of The Electrochemical Society, Inc. held in Cincinnati, Ohio.
8. Brors, D.L., Fair, J.A., Monnig, K.A., and Saraswat, K.C., Properties of low pressure CVD tungsten silicide as related to IC process requirements. *Solid State Technology,* April 1983, p. 183.
9. Inoue, S., Toyokura, N., Nakamura, T., Maeda, M., and Takaji, M., Properties of molybdenum silicide film deposited by chemical vapor deposition. *J. Electrochem. Soc.* 130:1603 (1983).
10. Gaczi, P. To be published.
11. Lehrer, W.I., Pierce, J.M., Good, E., and Justi, S., Low temperature LPCVD deposition of tantalum silicide, in *VLSI Science and Technology/1982,* eds. C.J. Dell'Oca and W.M. Bullis (The Electrochemical Society Pennington, NJ, Vol. 82-7), p. 258.
12. Wieczorek, C., Chemical vapor deposition of tantalum disilicide. *Thin Solid Films* 126:227 (1985).
13. Reynolds, G.J., Low-pressure chemical vapor deposition of tantalum silicide. To be published.
14. Tedrow, P., Ilderem, V., and Reif, R., Low pressure chemical vapor deposition of titanium silicide. *Appl. Phys. Lett.* 46(2):189 (1985).
15. Berkeley, J.F., Brenner, A. and Reid, W.E., Jr., Vapor deposition of tungsten by hydrogen reduction of tungsten hexafluoride. *J. Electrochem. Soc.* 114:561 (1967).
16. *Tungsten and Other Refractory Metals for VLSI Applications,* ed. R.S. Blewer (Materials Research Society, Pittsburgh, PA, 1986).
17. Lo, J-S, Haskel, R.W., Byrne, J.G. and Sosin, A., A CVD study of the tungsten-silicon system, in *Proceedings of the Fourth International CVD Conference,* eds. G.F. Wakefield & J.M. Blocker, Jr., Electrochem. Soc., Pennington, NJ (1973).

18. Shaw, J.M. and Amick, J.A., Vapor-deposited tungsten as a metallization and interconnection material for silicon devices. *RCA Review* 31:306 (1970).
19. Brors, D.L., Monnig, K.A., Fair, J.A., Coney, W., and Saraswat, K.C., CVD tungsten–A solution for the poor step coverage and high contact resistance of aluminum, *Solid State Technology,* April 1984, p. 313.
20. Smith, G.C., CVD tungsten contact plugs by in-situ deposition and etchback, *Proceedings of the Second International IEEE VLSI Multi-Level Interconnection Conference* (1985).
21. Miller, N.E. and Beinglass, I., Hot-wall CVD tungsten for VLSI. *Solid State Technology,* December 1980.
22. Lehrer, W.I. and Pierce, J.M., Low temperature CVD growth of tungsten disilicide, in *Semiconductor Silicon, 1981,* ed. H.R. Huff, R.J. Kriegler and Y. Takeishi, Electrochem. Soc., Pennington, NJ (1981).
23. Vogt, G.J., Low-temperature chemical vapor deposition of tungsten from tungsten hexacarbonyl. *J. Vac. Sci. Technol.* 20:1336 (1982).
24. Broadbent, E.K. and Ramiller, C.L., Selective low pressure chemical vapor deposition of tungsten. *J. Electrochem. Soc.* 131:1427 (1984).
25. Gargini, P., Tungsten barrier eliminates VLSI circuit shorts. *Ind. Res. & Dev.,* March 1983, p. 141.
26. Green, M.L. and Levy, R.A., Structure of selective low pressure chemically vapor-deposited films of tungsten. *J. Electrochem. Soc.* 132: 1243 (1985).
27. Broadbent, E.K. and Stacy, W.T., Selective tungsten processing by low pressure CVD. *Solid State Technol.,* Dec. 1985, p. 51.
28. Stacy, W.T., Broadbent, E.K. and Norcott, M.H., Interfacial structure of tungsten layers formed by selective low pressure chemical vapor deposition. *J. Electrochem. Soc.* 132:444 (1985).
29. Itoh, H., Nakata, R. and Moriya, T., Creep-up phenomena in tungsten selective CVD and their application to VLSI technologies. *IEDM Technical Digest,* International Electron Devices Meeting, IEEE (1985).
30. Wilson, R.H., Stoll, R.W. and Calacone, M.A., Highly selective high rate W deposition for via filling. *Proceedings of the Second International IEEE VLSI Multilevel Interconnection Conference* (1985), p. 343.
31. Cooke, M.J., Heinecke, R.A., Stern, R.C. and Maes, J.N.C., LPCVD of aluminum and Al-Si alloys for semiconductor application. *Solid State Technology,* Dec. 1982, p. 62.
32. Green, M.L., Levy, R.A., Nuzzo, R.G. and Coleman, E., Aluminum films prepared by metal-organic low pressure chemical vapor deposition. *Thin Solid Films* 114:362 (1984).

5

Plasma-Enhanced CVD

5.1 INTRODUCTION

In the previous two chapters, we examined the chemical vapor deposition process when well-defined reactant gases were supplied to the chamber that held the wafer. In all these cases, it was necessary to bring the wafer to a high temperature (350° to 900°C) before the desired reaction could be achieved. When integrated circuit dimensions are reduced (i.e., VLSI), diffused regions become quite thin (~2000 Å), and high-temperature processing is a disadvantage. Also, if we want to use aluminum in a multilevel metallization scheme, we have to keep temperatures below 500°C.

In such cases, another CVD deposition technique has been sought. The solution for some applications has been the use of a glow discharge. If the reactant gases in the reactor chamber are kept at low pressure (<2 Torr), then a glow discharge can be sustained, as discussed in Chapter 2. In a glow discharge, the entering gases are dissociated so that the reactive species that reach the wafer surface are atomic or molecular fractions. For example, a glow discharge in SiH_4 will produce SiH, SiH_2, or SiH_3 free radicals. All of these will react with unit probability on a wafer surface. Thus, it is not necessary to have the wafer at a high temperature in order to achieve acceptable deposition rates.

It may still be desirable to have a moderately high wafer temperature for other reasons. For example, at very low temperatures, film density may be low, or temperature may play an important role in determining film structure. Nonetheless, it is possible to operate at lower wafer temperatures than would be allowed by a strictly thermal process.

In the present chapter, we will review the nature of plasma-enhanced CVD (PECVD) films for a variety of applications. We will look at dielectrics (silicon nitride, silicon dioxide), semiconductors (polysilicon, epi silicon) and metals (refractory metals, refractory metal silicides, aluminum). There are many other important films (i.e., amorphous silicon for solar cells and TiN for tool hardening, among others) that have been put down by PECVD but they fall outside the scope of this text.

5.2 SILICON NITRIDE

PECVD of silicon nitride has been of commercial importance since 1976.[1] The original motivation was to find a final passivation layer for an integrated circuit that would replace the doped silicon dioxide films then in use. The latter were not reliable enough to permit packaging of integrated circuits in plastic. Silicon nitride was recognized as a better final passivation film, but the only available technique for its deposition was the high-temperature thermal process. Since it had to cover an aluminum final metallization layer that would melt at 600°C, this clearly could not work. The solution was to use PECVD at 350° to 400°C.

The reactant gases that have been used for PECVD of silicon nitride have been either $SiH_4 + NH_3$ with a diluent gas such as N_2, Ar or He, or $SiH_4 + N_2$. Reactors using parallel-plate, capacitively-coupled RF have generally been employed (see Chapter 2). The specific reactor geometry or condition (hot or cold wall) does not have a major influence on the film characteristics. It does, however, play a large role in determining film deposition rate and uniformity, which for batch systems determines the commercial viability of the reactor.

After we evaluate the nature of the silicon nitride deposited by PECVD, we will return briefly to the question of "characterizing" a production reactor.

Once the reactor configuration has been selected, some of the parameters that will determine film quality are:

(1) operating pressure

(2) operating temperature

(3) discharge frequency

(4) reactant gas mixture

PECVD silicon nitride films are amorphous at the temperature used for growth, so film structure (i.e., grain size) is not an issue. The film quality is determined by:

(1) stoichiometry

(2) H_2 content

(3) impurities

(4) density

(5) stress

The first three items relate to the chemical nature of the film. An outstanding feature of PECVD silicon nitride films is that their stoichiometry can be controlled, and that they can have as much as 30 atomic percent hydrogen in them. The last two items relate to the mechanical behavior of such films. If they are not dense enough, they will not be effective barriers to moisture and oxygen. Films with tensile stresses tend to crack, and ones with strong compressive stresses tend to delaminate readily.

For silicon nitride films made in a cold-wall, parallel-plate reactor operating at 50 kHz, 200 mTorr, gas flows of $SiH_4/NH_3/N_2$ = 140/270/800 sccm and 500 watts of power, we can compare chemical and physical properties with thermally-deposited silicon nitride. Such a comparison is shown in Table 1.[2]

Table 1: Physical and Chemical Properties of CVD and PECVD Silicon Nitride Films[2]

Property		High Temp. Nitride 900°C	Plasma Dep. Nitride 300°C
Composition		Si_3N_4	SiN_x
Si/N ratio		0.75	0.8–1.0
Solution Etch Rate			
Buffered HF	20°–25°C	10–15 Å/min	200-300 Å/min
49% HF	23°C	80 Å/min	1500-3000 Å/min
85% H_3PO_4	155°C	15 Å/min	100-200 Å/min
85% H_3PO_4	180°C	120 Å/min	600–1000 Å/min
Plasma Etch Rate			
92% CF_4-8% O_2, 700 W		600 Å/min	1000 Å/min
Na^+ penetration		<100 Å	<100 Å
IR Absorption			
Si-N max.		~830 cm^{-1}	~830 cm^{-1}
Si-H minor		–	2,200 cm^{-1}
Density		2.8-3.1 g/cm^3	2.5-2.8 g/cm^3
Refractive Index		2.0-2.1	2.0-2.1
Dielectric Constant		6–7	6–9
Dielectric Strength		1×10^7 V/cm	6×10^6 V/cm
Bulk Resistivity		10^{15}–10^{17} Ω-cm	10^{15} Ω-cm
Surface Resistivity		$>10^{13}$ Ω/sq	1×10^{13} Ω/sq
Intrinsic Stress		1.2–1.8×10^{10} dyn/cm^2 Tensile	1-8×10^9 dyn/cm^2 Compressive
Thermal Expansion		4×10^{-6}/°C	–
Color, Transmitted		None	Yellow
Step Coverage		Good	Conformal
H_2O Permeability		Zero	Low-None

If we restrict our attention to one set of chemical precursors (SiH_4, NH_3, N_2), we have available more detailed data describing the quality of PECVD silicon nitride thin films as a function of the several operating parameters.[3] Experiments were carried out in a parallel-plate reactor placed in a horizontal hot tube system where the wafer was placed on the grounded electrode.

We can see the contribution of relative NH_3 concentration on film composition (N_A at %), density (ρ g/cm^2) and deposition rate (G nm/min) for a 300°C, 310-kHz and 65-Pa (487.5 mTorr) deposition in Figure 1. For this case, the SiH_4 flow was 100 sccm and the $N_2 + NH_3$ flow was 1,400 sccm.

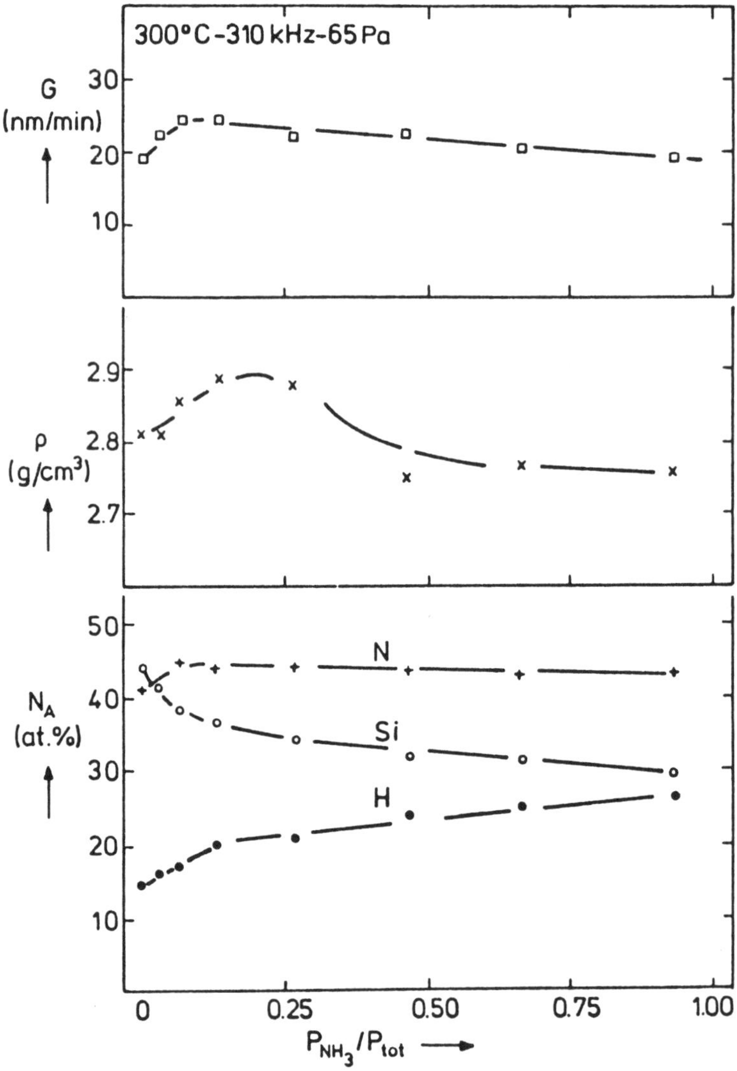

Figure 1: Silicon nitride film characteristics as a function of relative NH_3 concentration in reactant flow.[3] Reprinted by permission of the publisher, The Electrochemical Society, Inc.

We observe that as the partial pressure of NH_3 is increased, the amount of hydrogen incorporated into the film rises from 14 to 26 atomic %. The film density is a maximum where the Si/N ratio is 0.75.

Similar measurements were made as temperature varied from 300° to 600°C at 130 Pa (975 mTorr) and flows set at SiH_4 = 100 sccm/N_2 = 200 sccm/NH_3 =

1,200 sccm with a discharge frequency of 310 kHz. For comparison, results at 700°C without a discharge were included and are shown in Figure 2. As the temperature increases, the density rises to a limiting value and the hydrogen content of the film drops rapidly to less than 10%. It is interesting to note that "thermal" silicon nitride still has several percent hydrogen in the film.

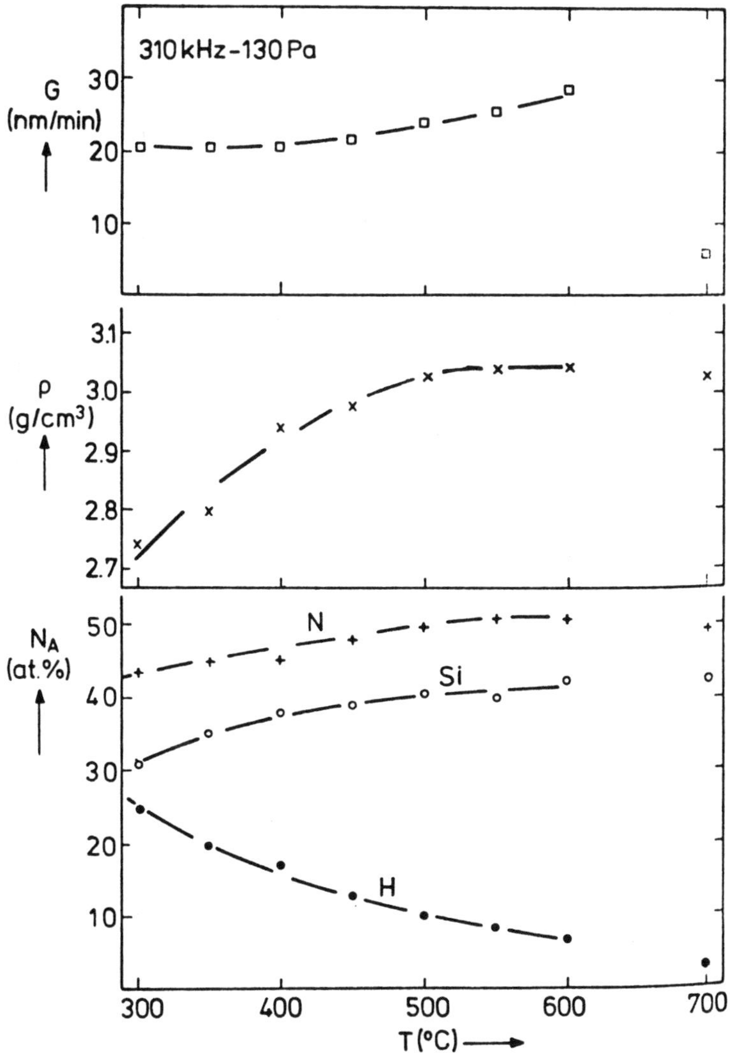

Figure 2: Silicon nitride film characteristics as a function of deposition temperatures.[3] Reprinted by permission of the publisher, The Electrochemical Society, Inc.

Variations with pressure for a 300°C, 310-kHz deposition with SiH_4 = 100 sccm/N_2 = 300 sccm/NH_3 = 1,100 sccm are shown in Figure 3. Most notably here, the hydrogen content rises and the film density drops sharply as pressure is increased above 150 Pa (1.125 Torr).

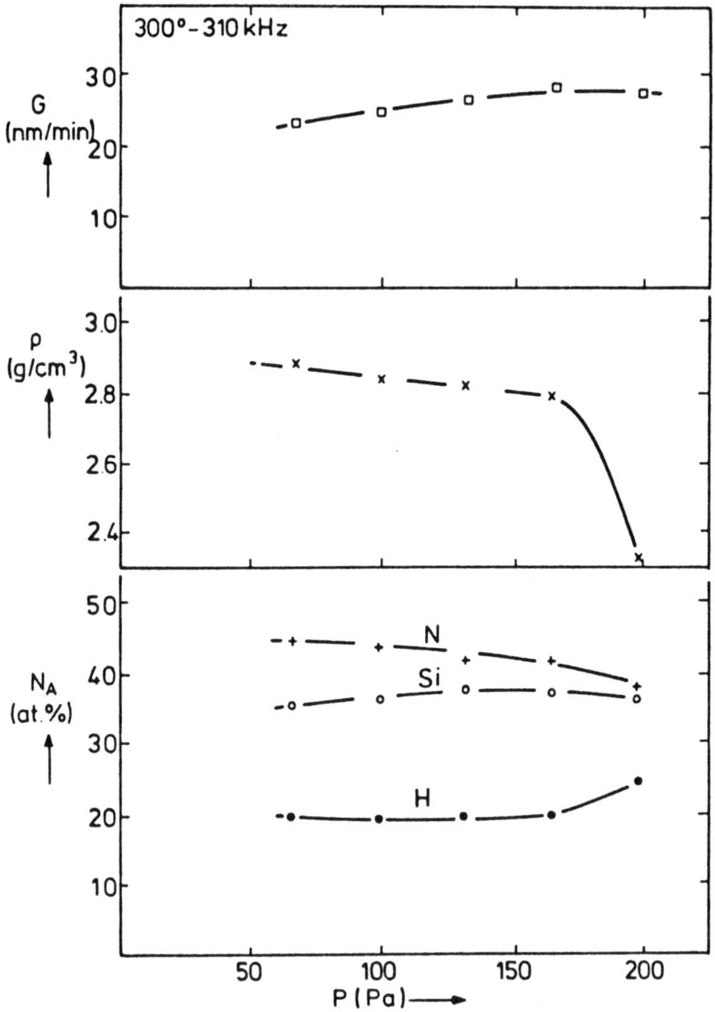

Figure 3: Silicon nitride film characteristics as a function of pressure.[3] Reprinted by permission of the publisher, The Electrochemical Society, Inc.

A fairly dramatic change in film characteristics is seen as frequency is varied. For a 300°C, 130 Pa (975 mTorr) discharge with SiH_4 = 100 sccm/N_2 = 700 sccm/NH_3 = 700 sccm, the variation with frequency is shown in Figure 4. At frequencies of 4 MHz and above, the film density drops sharply and the hydrogen content rises sharply.

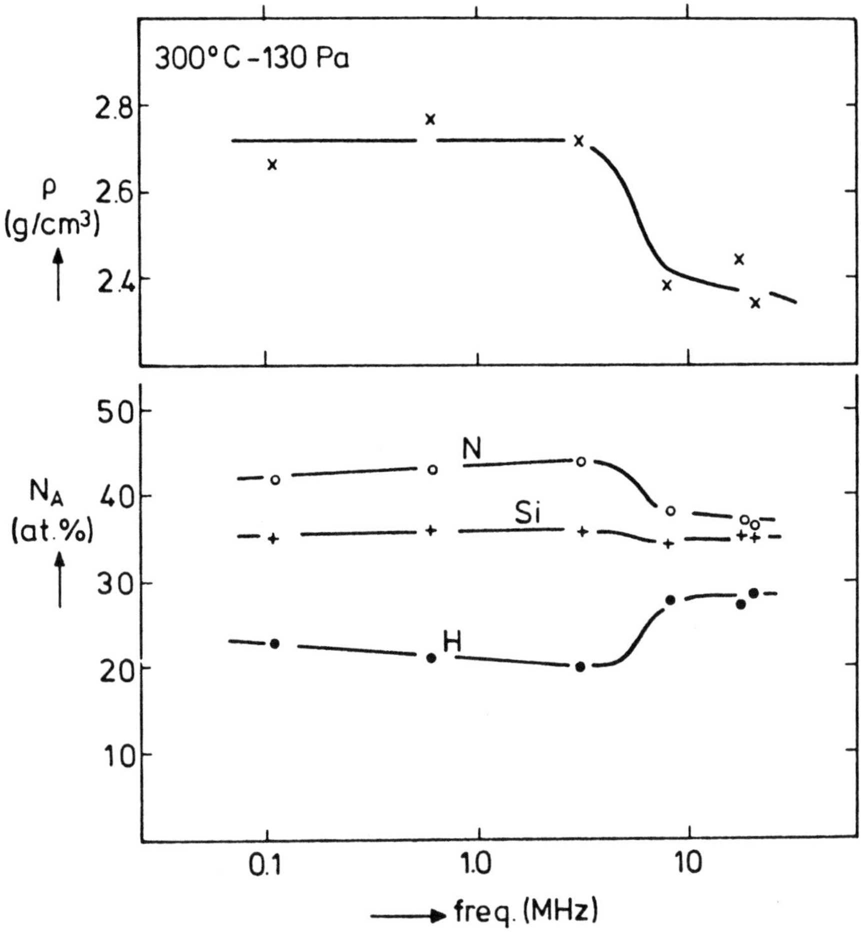

Figure 4: Silicon nitride film characteristics as a function of frequency.[3] Reprinted by permission of the publisher, The Electrochemical Society, Inc.

Finally, we can see how the film stress behaves with pressure, temperature and frequency in Figure 5. For each variable, the flow conditions are the same as those specified in the corresponding prior figure. Clearly, depositions at lower frequencies and temperatures tend to favor the desirable compressive stress condition.

As noted in Chapter 2, when discharge frequency is low, there is a greater tendency toward ion bombardment. If we argue that ion bombardment compresses the film as it is being deposited, we would expect to see compressive stresses and higher density films, as shown in Figures 4 and 5. Attempts to minimize the H_2 content of the films is best done by operating with less NH_3 and at higher temperatures.

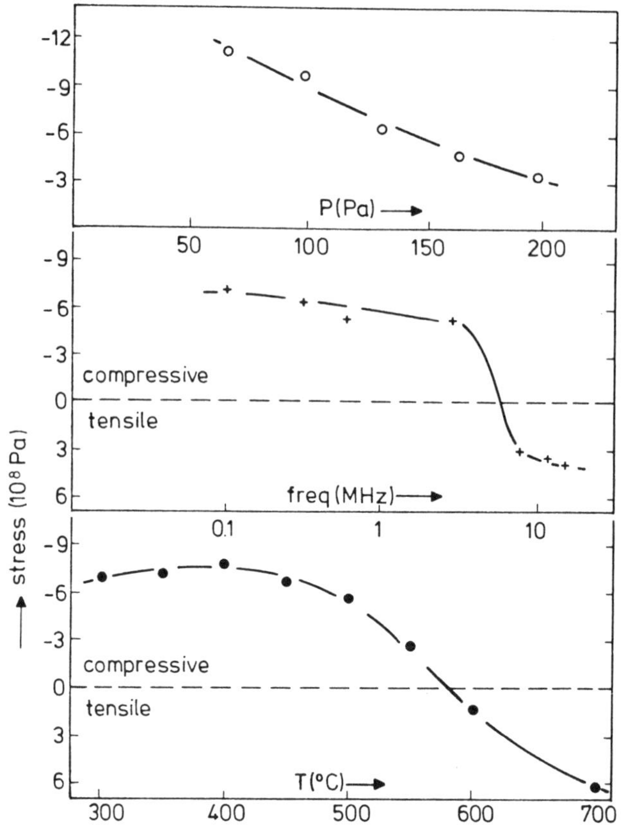

Figure 5: Stress in silicon nitride films as functions of pressure, frequency and temperature.[3] Reprinted by permission of the publisher, The Electrochemical Society, Inc.

As would be expected, there are some differences in the films formed when different diluent gases are used (i.e., Ar or H_2 rather than N_2). Recent experiments have been carried out in a parallel-plate, cold-wall PECVD reactor[4] operating at 300°C, p = 0.33 mbar (330 mTorr), an RF power (50 kHz) of 200 W, and a total gas flow of 1.2 slm. The ratio of NH_3 to N_2, Ar or H_2 was varied while maintaining the sum of the NH_3 and the diluent constant. In plasma silicon nitride films, the composition can vary widely depending on deposition conditions. When we speak of film composition, we are talking about the Si/N ratio, which would ideally be 0.75, and the H_2 content. These authors demonstrate that the film refractive index is directly related to the Si/N ratio, as shown in Figure 6. Then, when the fraction of NH_3 is varied during deposition, we can see in Figure 7 that the stoichiometry varies differently with NH_3 concentration, depending on the diluent used.. For example, to achieve a Si/N ratio of 0.85, Figure 6 says we need a refractive index of 2.0. For the higher SiH_4 flow in Figure 7, we need only 35% NH_3 in the reactant gases when we use N_2 as compared to 53% if we use argon as the diluent.

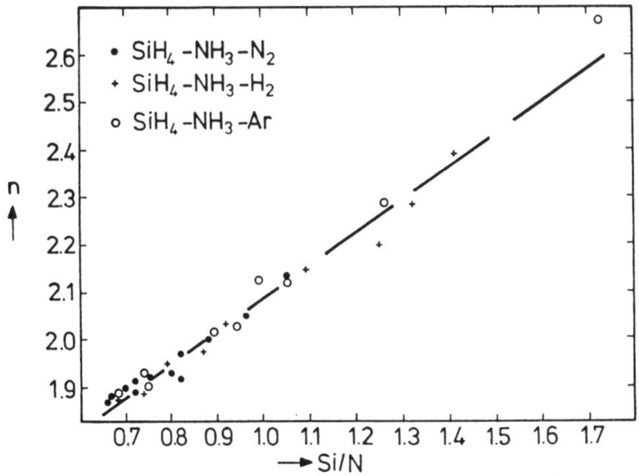

Figure 6: Refractive index of silicon nitride films as a function of stoichiometry (Si/N).[4] Reprinted by permission of the publisher, The Electrochemical Society, Inc.

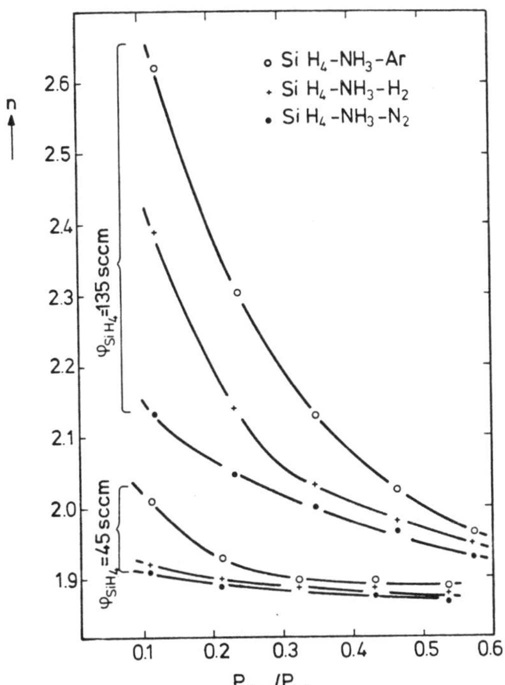

Figure 7: Refractive index of silicon nitride films versus ammonia concentration for different diluents.[4] Reprinted by permission of the publisher, The Electrochemical Society, Inc.

As diluents are changed, there does not appear to be any change in hydrogen content of the films. However, the film density does show some dependence on diluent used, as shown in Figure 8. Evidently, the densest films are deposited when argon is used as the diluent.

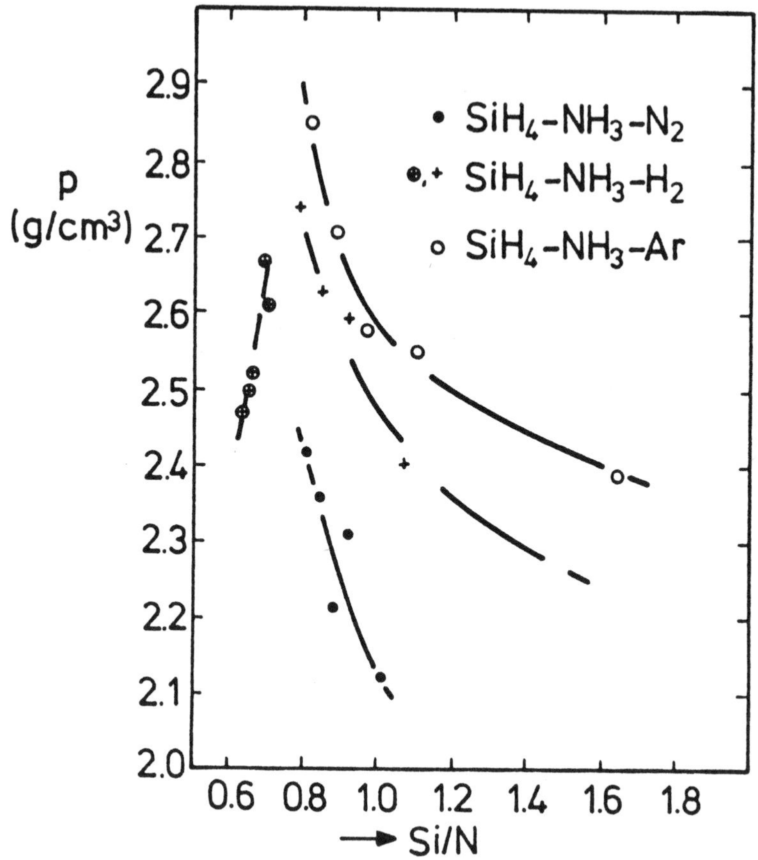

Figure 8: Silicon nitride film density as a function of Si/N ratio for different diluents.[4] Reprinted by permission of the publisher, The Electrochemical Society, Inc.

When NH_3 is omitted from the reactant gases and the N in Si_3N_4 is provided by N_2, substantial changes occur in the films deposited. The % H incorporated in such films is much less than when NH_3 is used, for the same deposition temperature as shown in Figure 9.[5]

It is also interesting to note that the % H in silicon nitride films varies widely depending on the type of reactor used (15.4 to 30.7 at. %).[5] It is not clear that the hydrogen content of these films is undesirable in a plasma silicon nitride film. However, if for some reason the decision is made to reduce H

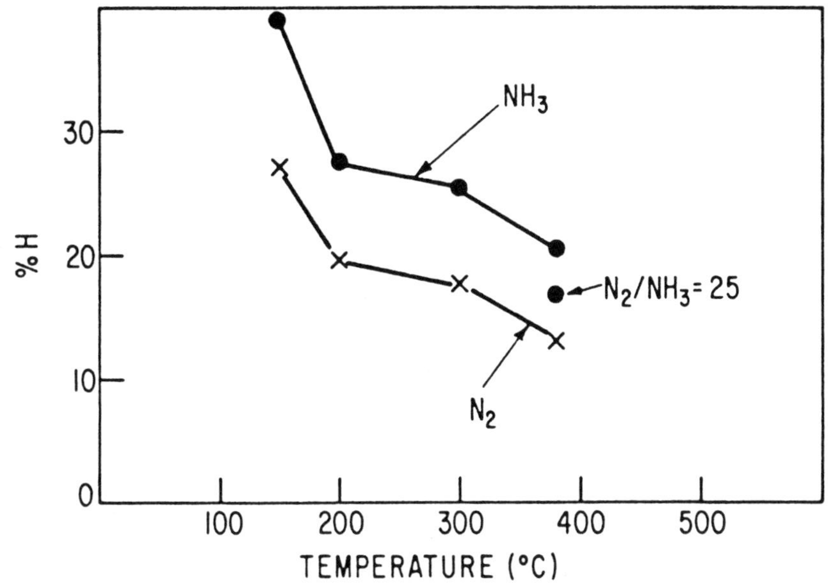

Figure 9: Hydrogen % versus deposition temperature for films deposited with $SiH_4 + NH_3$, $SiH_4 + N_2$ and $SiH_4 + NH_3 + N_2$ (one point).[5]

in the film, it would be best to work with a $SiH_4 + N_2$ process at as high a temperature as possible. As noted earlier in Chapter 3, however, even thermally-deposited, low-pressure CVD films retain some hydrogen in the film.

Earlier, we reviewed silicon dioxide (thermal) films deposited with added phosphorus to serve as a getter for mobile ion impurities, as a final passivation film. Plasma-enhanced silicon nitride can also be doped with phosphorus.[6] Some of the film characteristics have been reviewed, and it was found that the films with 2 to 3% P had the best electrical quality. No measurements of stress or H_2 content were reported, so it is not clear that these would be useable films.

It is also interesting to note that the H bound into the plasma silicon nitride film can be driven off, to some extent, by annealing at low temperatures. Data for $SiH_4 + N_2 + He$ and $SiH_4 + N_2 + Ar$ depositions are shown in Figure 10.[7] Observe that for a film deposited with He at 300°C and then annealed for 10 hours at 300°C, the hydrogen content drops from 19 to 12%. The curves marked NOT are unannealed. Depositions with Ar did not display this behavior.

After the quality of the plasma silicon nitride films and their dependence on the several system parameters has been evaluated, there still remains the question of whether or not a given process can be commercially viable. Here the issue is the deposition rate and the uniformity of deposition on a wafer and over all wafers in the reactor. The ideal solution is to deposit at a high rate uniformly over many wafers at one time. We cannot simply stack many wafers close together and run a low-pressure process, as in thermal LPCVD, because we have to be sure the plasma discharge is uniform as well.

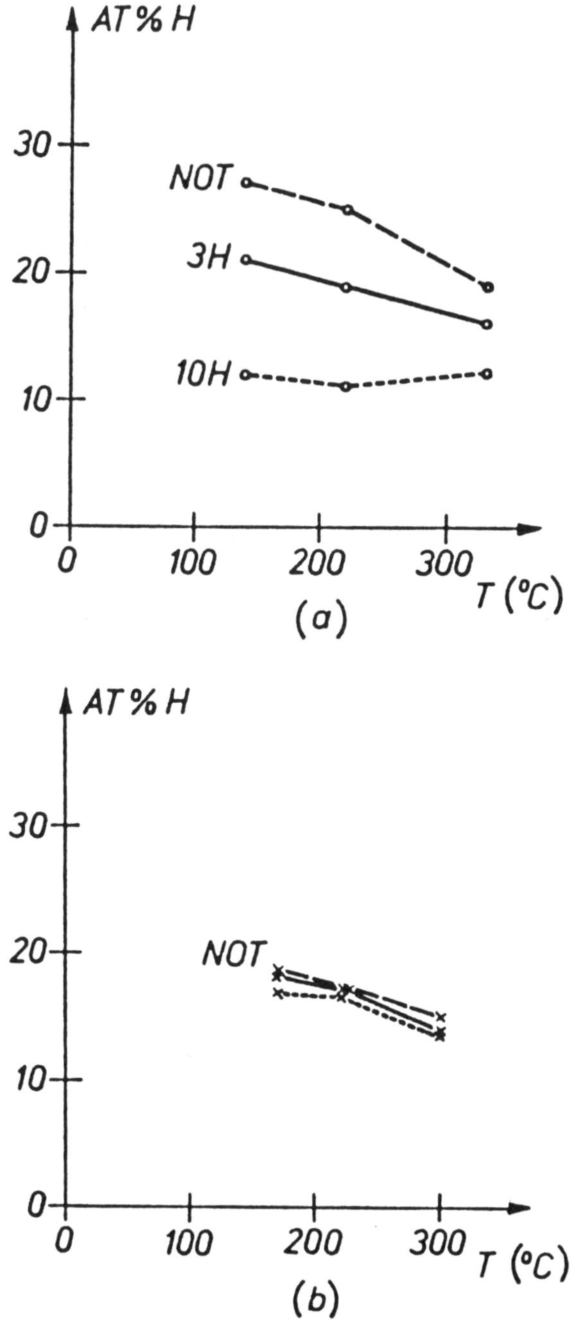

Figure 10: Effect of annealing on hydrogen % in plasma silicon nitride films versus deposition temperature: (a) He diluent; (b) Ar diluent.[7] Reprinted by permission of the publisher, The Electrochemical Society, Inc.

In fact, there are many factors that make a reactor "commercial" or production ready, and they will be discussed along with many examples in Chapter 6. In the present chapter, we will just touch on the typical process modifications necessary to develop a "commercial" system.

The process of "characterizing" a reactor can be illustrated for a parallel-plate cold-wall reactor operated at 50 kHz.[8] System power was kept at 500 W, pressure at 200 mTorr, and wafer temperature of 240°C. Wafers are placed on a circular electrode which is rotated to promote uniformity of deposition. Therefore, we are only interested in the radial variation of deposition rate. Reactive gases enter at the center and flow out at the periphery.

Typically, deposition uniformity depends on the reactive gas mixtures, as is illustrated in Figure 11.[1] Clearly, much trial and error is required to establish reactor operating conditions that will yield wafers with uniform deposits.

It is interesting to observe that, as shown in Figure 11(a), as the SiH_4 flow is decreased, the deposition rate at the outer portions of the platen rise rapidly. Running with N_2 and no NH_3, as in Figure 11(d) makes it impossible to obtain uniform depositions.

There is no convenient, obvious, explanation for these diverse results. Basically, as we change the gas composition, we are modifying the glow discharge behavior even though power and pressure have remained unchanged. Such changes in the discharge are the cause of unexpectedly nonuniform deposition behavior. Since no theoretical description of such a complex system is available, reactor uniformity must be obtained by trial and error, or "characterization." As experience with such systems grows, it should eventually be possible to model them and make theoretical predictions of their performance.

5.3 SILICON DIOXIDE AND OXYNITRIDES

The only other plasma-enhanced CVD film that has seen wide use in integrated circuit manufacture is the plasma oxide film. We say "so-called" because it is not truly SiO_2, but rather $SiO_xN_yH_z$. In fact, it is just this ability to modify the film stoichiometry that makes these films so valuable. Many of the film characteristics change depending on this stoichiometry, so it allows a freedom to alter film characteristics that is not possible with thermally-grown films.

Plasma oxide can be grown from a number of oxidizers plus SiH_4. Among these are N_2O, O_2, CO_2 and even TEOS (tetraethoxysilane). Generally, O_2 is not used as it too often leads to homogeneous nucleation. The preferred reactants have proven to be SiH_4 and N_2O, so we will restrict our discussion to these. Films grown in both cold-wall and hot-wall reactors will be considered.

A very thorough study of plasma oxides deposited in a parallel-plate cold-wall system has been reported by van de Ven.[8] The operating frequency used was 57 kHz, the pressure was 400 mTorr, and the wafers were held at 300°C. The influence of the N_2O/SiH_4 ratio, and the power level are shown in Figure 12. As expected, deposition rates increase with power level. The etch rate in buffered oxide etch (BOE) also increases, suggesting a less dense film with the higher deposition rate. Higher N_2O/SiH_4 ratios give a lower index of refrac-

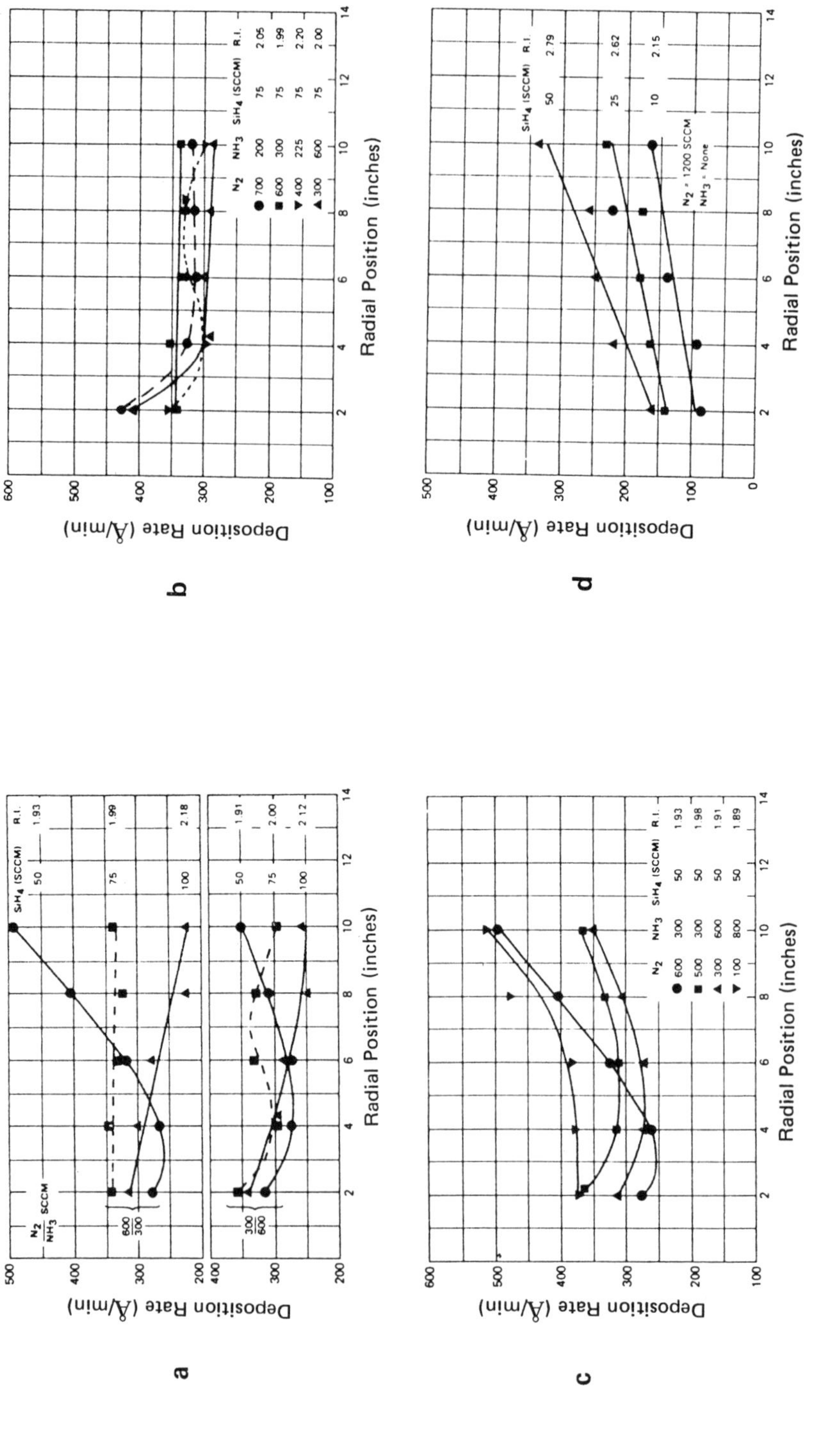

Figure 11: Deposition rates versus reactive gas flows.[1]

tion, which approaches the value appropriate for thermal oxide of 1.48. In general, higher values of N_2O/SiH_4 and lower power levels are preferred to minimize gas phase reaction and nucleation.

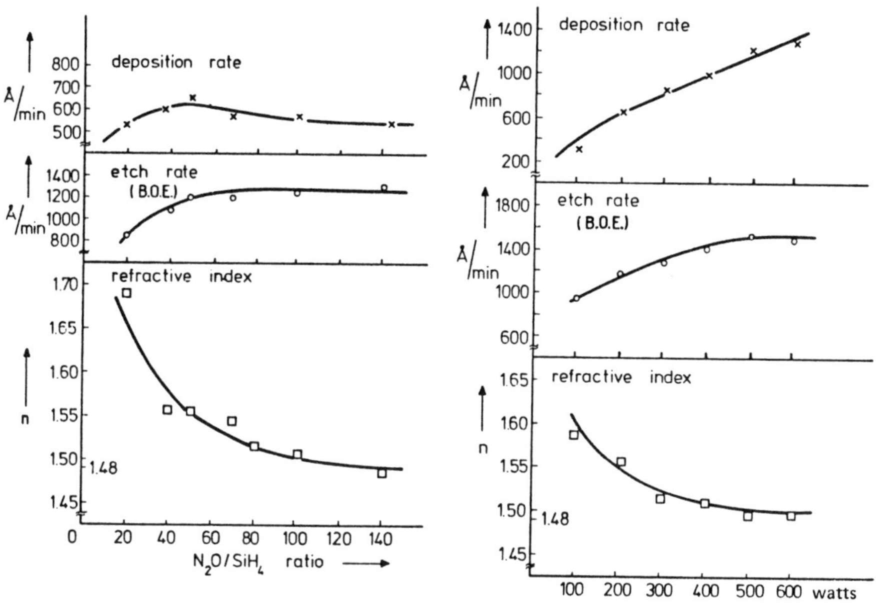

Figure 12: Film characteristics for plasma oxide deposition versus gas composition and power level.[8]

Results obtained for a typical process condition are shown in Table 2, where plasma oxide and nitride films are compared. In spite of the power level being a third that used for plasma nitride, the plasma oxide deposition rate is almost twice as high (~600 Å/min). This is just another indication of the ease with which silane can be oxidized compared to its being nitrided. This data indicates only a small quantity of N in the films. However, a 2 to 3% level of hydrogen is also mentioned.[8] Other experiments carried out at 13.56 MHz and 1 Torr pressure[9] show hydrogen content as high as 8% for some films. It would appear that the higher degree of ionization (and dissociation) achievable at the higher frequency stimulates hydrogen incorporation. On the other hand, the stress apparently remains compressive even at the higher frequency.[9]

Plasma oxide has found utility in high-frequency applications for dual-layer isolation,[8] because of its low dielectric constant and high breakdown voltage. Also, it is in compression when deposited, so that it can be used as the dielectric when thick films (2 to 5 microns) are needed. Such thick films when deposited by thermal CVD (which is deposited in tension) tend to crack. One final advantage to the use of plasma oxide rather than plasma nitride is that

Table 2: Plasma Oxide and Nitride Characteristics[8]

	Silicon Dioxide	Silicon Nitride
Gases	$SiH_4 + N_2O$	$SiH_4 + NH_3 + N_2$
% SiH_4	2%	9%
% N_2O, NH_3 resp.	98%	45%
RF Power Density	0.05 W/cm²	0.17 W/cm²
RF Frequency	57 kHz	57 kHz
Operating Pressure	53 Pa	33 Pa
Substrate Temperature	300°C	300°C
Deposition Rate	60 nm/min	38 nm/min
Film Uniformity	±5%	±4%
Film Composition	$SiO_{1.9}N_{0.15}$	$Si_{3.2}N_4$ (H)
Refractive Index	1.54	2.02
Film Density	2.38 g/cm³	2.75 g/cm³
Etch Rate (B.O.E.)	130 nm/min	20 nm/min
Etch Rate ($CF_4 + O_2$ plasma)	10 nm/min	150 nm/min

the hydrogen content can be much lower (i.e., 2 to 3% versus 20 to 30%). For applications where evolution of H_2 from a plasma nitride layer during later high-temperature processing would be deleterious, the plasma oxide is preferred.

As noted earlier, the stoichiometry of plasma oxide films can be adjusted by changing deposition conditions. The electrical behavior of a composite film consisting of a thin thermal oxide covered by a thin, silicon-rich, plasma oxide has been studied.[10] The silicon-rich film actually consists of silicon crystals interspersed within the plasma oxide. It is deposited in a 13.56 MHz, parallel-plate, cold-wall reactor operated at 600 mTorr with the wafers at 350°C. The ratio of N_2O flow to SiH_4 flow was varied during the experiments from 0 to 150 to alter the stoichiometry. As just one illustration of film behavior (composite film in this case), we show the dielectric constant as a function of N_2O/SiH_4 flow in Figure 13.[10] Increasing the SiH_4 flow and thus increasing the silicon excess leads to a substantial increase in dielectric constant. Apparently, the high dielectric constant of the Si-rich plasma oxide films is used for dual-dielectric storage capacitors in dynamic memories.

If desired, plasma oxide films can be doped much as the plasma nitride film we discussed earlier. In fact, doping with boron and phosphorus has been carried out as an alternative to the standard atmospheric-pressure thermal CVD process for BPSG.[11,12] The latter process has the drawbacks of high defect density and poor thickness uniformity, so it was hoped that plasma BPSG would be an improvement. However, there are differences in the films in terms of H_2 and N_2 content, and their effect on reflow temperature, intrinsic stress and passivation effectiveness had to be examined.

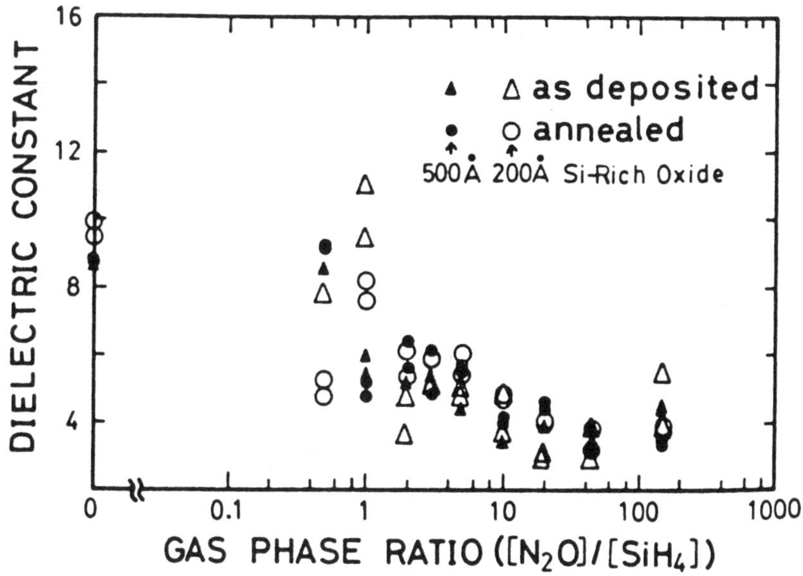

Figure 13: Dielectric constant as a function of flow ratio (N_2O/SiH_4).[10]

Depositions were done in a hot-wall tubular plasma reactor at a frequency of 410 kHz. Deposition temperatures ranged from 300° to 400°C, and pressures from 830 mTorr to 1.5 Torr. Gases used were SiH_4, N_2O, O_2, PH_3 and B_2H_6, with the latter two diluted by argon. The boron and phosphorus concentrations were adjusted by changing the reactant gas mixture.

Stress in the as-deposited films was compressive, as expected. After densification (800°C for 1 hour in steam), the stress was reduced, and in some cases made slightly tensile. Comparison of the plasma oxide films with thermal oxide films is shown in Table 3. Given the compressive stress found in the plasma oxide films, we would expect improved crack resistance. Tests were carried out[11] on a 1-micron plasma BPSG film (P = 3.4 wt %, B = 3.2 wt %) over 1-micron thick aluminum lines by thermal stressing at 460°C. There were no microcracks, although thermal atmospheric pressure BPSG would crack under the same conditions. No major differences were found in step coverage or reflow characteristics, as compared to the thermal films.

The ability of the plasma BPSG film to passivate against sodium penetration was compared to the thermal film. Evidence was found of some sodium penetration in the plasma films, and none in the thermal ones.

Since one major motivation behind the use of plasma BPSG was to provide an improved passivation barrier, the better crack resistance is an advantage, but the greater sodium penetration is a negative. Therefore, it is not clear if it would be advantageous to make this replacement[11] for a final passivation film. Its use as an inter-metallic dielectric may be more useful.

Table 3: Comparison of Stress in Plasma and Thermal Oxide Films[11]

Glass Type	Stress (× 10^{-9} dynes cm^{-2}) As-Deposited	Densified
Plasma Glass (Non-Doped)	-3.0	2.9
Plasma PSG (10%)	-1.66 (C)	-
Plasma BPSG	-(1-2) (C)	(0.7) to -1
Thermal Oxide	-(3-4) (C)	-
LPCVD PSG High Temp. Dep.	-0.5 (C)	-
APCVD PSG (Low Temp.)	+(1-4) (T)	-

C = Compressive
T = Tensile

Finally, we will consider PECVD silicon oxynitrides, and their unique characteristics. When oxygen is added to a PECVD nitride film, there are indications that it may improve its crack resistance as a final passivation layer.[13] Also, there may be advantages in terms of its electrical characteristics as an interlayer dielectric. Therefore, the nature of films grown when N_2O is added to a SiH_4, NH_3 and He gas mixture in a high frequency (13.56 MHz), cold-wall, parallel-plate reactor have been studied.

Measurements were made showing that the film composition could be varied uniformly from nitride to oxide, with varying composition oxynitrides in between, simply by changing the ratios of reactive gases appropriately. The refractive index of the films, accordingly, varied from 1.93 (nitride) to 1.45 (oxide). Annealing did not alter the composition, except to reduce the hydrogen content. Increasing the oxygen content of the films had the effect of decreasing the plasma etch rate of the film.

The ability to deposit oxynitride films at low temperatures by PECVD is just another example of the unique ability of this technique to tailor the chemical composition of a film to match desired film characteristics.

5.4 POLYSILICON

Up to this point, we have been considering PECVD films that are being used commercially in the manufacture of integrated circuits. We will now consider PECVD of other films that may find practical application in the future. At the present time, they are still considered experimental.

The major advantage to the use of PECVD for deposition of polysilicon films is the possibility of depositing them at low temperatures. The conventional LPCVD process requires T ~650°C for times on the order of a half hour. PECVD of polycrystalline films can be carried out with a wide variety of gases (SiH_4, Ar, He, H_2 and doped or undoped) at temperatures ranging from 200°

to 400°C.[14] The morphology of the films created is an important issue. In some cases, polycrystalline films are formed, while in others a mixture of microcrystals in an amorphous matrix is found.

Aside from the reactant gases that can be used, another factor influencing the nature of the films is ion bombardment. If ion bombardment is increased at a given deposition temperature, it is possible to convert an amorphous film to one which is microcrystalline. As ion bombardment continues to increase, the volume of crystallites display lower resistivity, as would be expected. The as-deposited resistivity of PECVD polysilicon has been shown to have a value of 0.78 x 10^6 Ω-cm, as compared to s similar film done by LPCVD which is 13.8 x 10^6 Ω-cm.

Many questions remain. It will have to be shown that the doped resistivity can be as good as conventional doped LPCVD films. Also, the role of species other than silicon has to be clarified (i.e., hydrogen, noble gases).

At the same time, the fully doped polysilicon resistivity (LPCVD or PECVD) is proving inadequate for future generation IC devices, and interest is shifting to refractory metal silicides and/or refractory metals as replacements. Therefore, it is not clear how significant the development of such a new process will be for integrated circuit manufacture. There are other applications, however, where it may be better suited.

5.5 EPITAXIAL SILICON

Referring to our earlier discussion of the thermal CVD of epitaxial silicon, we recall that a very high temperature initial etch (HCl at 1150° to 1250°C) was necessary to prepare the silicon substrate for epitaxial deposition. Such very high temperatures can, and do, lead to difficulties with slip defects. Tnen it was noted that commercial practice was to deposit from $SiCl_4$ or $SiHCl_3$ rather than SiH_2Cl_2 or SiH_4. The latter two reactants permit high deposition rates at lower temperatures, but in some cases, do not produce good quality films. Therefore, deposition temperatures range up to 1000° to 1050°C, and at these temperatures, autodoping becomes a serious problem, especially for the thinner epi films of current interest.

Accordingly, there has been considerable interest in carrying out this process at lower temperatures. Some research has been done using a PECVD approach to deposit epitaxial silicon films from SiH_4 at temperatures as low as 750°C without the high-temperature cleaning step.[15] Unfortunately, similar research has not been done using the reactants traditionally used. In fact, referring to Figure 17 of Chapter 4, we recognize that epi silicon from SiH_4 has always been possible at relatively low temperatures. The issue has been, that the film quality has been inadequate. It remains to be seen whether or not the PECVD approach can produce better quality films than can be produced thermally.

The research at MIT has been done in the cold-wall vertical tube reactor shown in Figure 14. The wafer is aligned almost parallel to the flow on a vertical silicon carbide-coated susceptor. The wafer is heated by optical radiation from high-intensity lamps to a temperature of 775°C. Silane was introduced at a pressure of 15 mTorr for the deposition. Prior to deposition, an argon plasma was struck and the wafer biased 300 V negatively. The argon ion bombardment effectively removed the approximately 20 Å native oxide from the wafer and permitted epi film growth even with the plasma off (~340 Å/min) at 775°C. Using the plasma during epi growth increased the deposition rate to 450 Å/min. If the plasma preclean is not used, or less than 300 V bias is applied, then films deposited are polycrystalline.

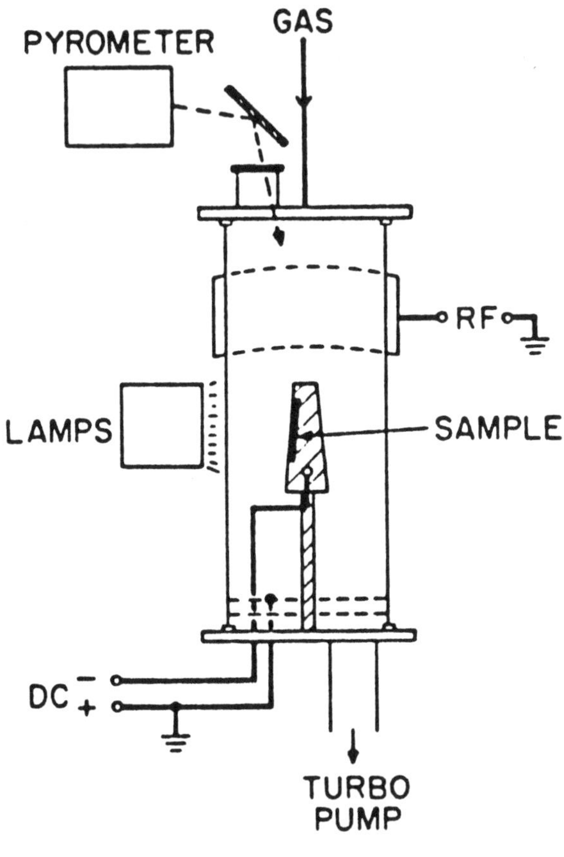

Figure 14: MIT system for PECVD of epi silicon.[15]

There is no doubt that the deposition rate for silicon deposition can be enhanced by plasma excitation of the reactive gases. The modest increase described above (340 → 450 Å/min) is probably more a function of the reactor configuration than anything basic in the process. Whether the epi films deposited at these higher rates are of good enough quality for integrated circuit fabrication remains to be seen. It would also appear that if the traditional high-temperature HCl and/or H_2 predeposition etch can be avoided by a plasma process, then process temperatures can probably be realistically reduced by 50° to 100°C. This would reduce slip defects and problems from autodoping.

5.6 REFRACTORY METALS AND SILICIDES

5.6.1 Tungsten

In the previous chapter, it was shown that it is possible to deposit quality tungsten films by CVD for either selective or blanket applications at moderate temperatures. As-deposited resistivities are low, so post-deposition high-temperature anneals are not essential.

As research on thermal CVD of tungsten proceeded, parallel studies were being done on plasma-enhanced depositions.[16] At the present, commercial application of thermal CVD tungsten has begun, but all the problems have not been solved. Therefore, it will be useful to review some of the PECVD of tungsten studies as well as similar research for other refractory metals or their silicides.

Tungsten films have been deposited in a parallel-plate, cold-wall reactor operating at 4.5 MHz. Initial depositions were attempted at temperatures from 40°C on up with WF_6 as the reactant gas. It was discovered that depositions could not be accomplished at temperatures above 90°C. Apparently, the glow discharge creates many free fluorine atoms, and their etching of tungsten overtakes any deposition from WF_6 decomposition on the surface at this temperature. The solution to higher temperature depositions was the addition of H_2 to the WF_6 to promote the formation of HF in the glow discharge. The reaction is similar to the surface reaction we discussed for blanket tungsten deposition in the last chapter, only now W is released in the gas phase.

$$\text{(1)} \qquad WF_6 + 3H_2 \rightleftharpoons W + 6HF.$$

In effect, the hydrogen serves to scavenge any free fluorine in the plasma, thereby promoting the deposition reaction.

Films deposited at low temperatures were porous and exhibited high resistivity. Therefore, depositions were carried out from 250° to 400°C and produced deposition rates that increased rapidly and then leveled off, as shown in Figure 15. Deposition rate increased linearly with pressure and power level.

The as-deposited resistivity of the PECVD tungsten was found to be quite high, as is shown in Figure 16. X-ray diffraction analysis[17] showed that the as-deposited film was largely a metastable β-phase of tungsten. Pure β-phase tungsten has a resistivity over 100 $\mu\Omega$-cm, whereas the material we have been considering all along has a low resistivity (~5 $\mu\Omega$-cm) and is referred to as the α-phase. Heat treatment of the as-deposited films, as shown in Figure 17, brings the tungsten resistivity back into the expected range. Analysis of the chemical composition of the PECVD tungsten films showed less than 1% fluorine incorporated.

If SiH_4 is added to the reactive gas mixture, then tungsten silicide can be deposited.[18] In this case, He was used as a diluent, along with the WF_6 and SiH_4, and deposition was carried out in a parallel-plate, cold-wall plasma

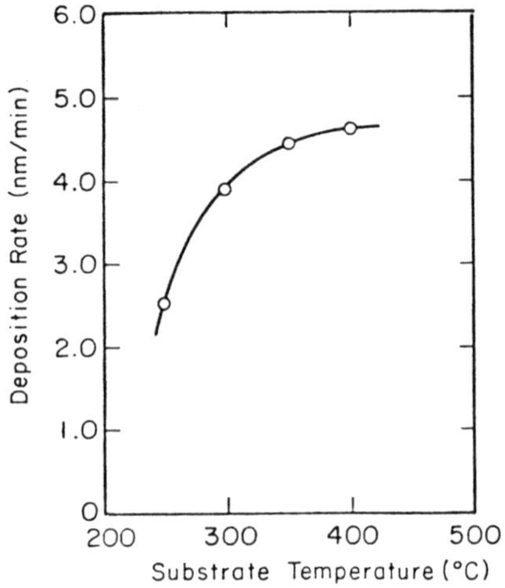

Figure 15: Tungsten deposition rates versus substrate temperature for p = 200 mTorr, $H_2/WF_6 = 3$ and 30 watts.[16]

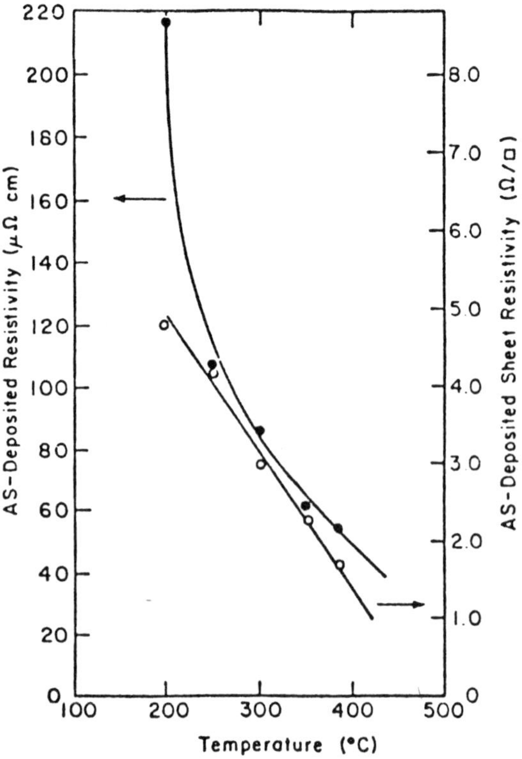

Figure 16: As-deposited resistivity and sheet resistance of PECVD tungsten versus deposition temperature.[16]

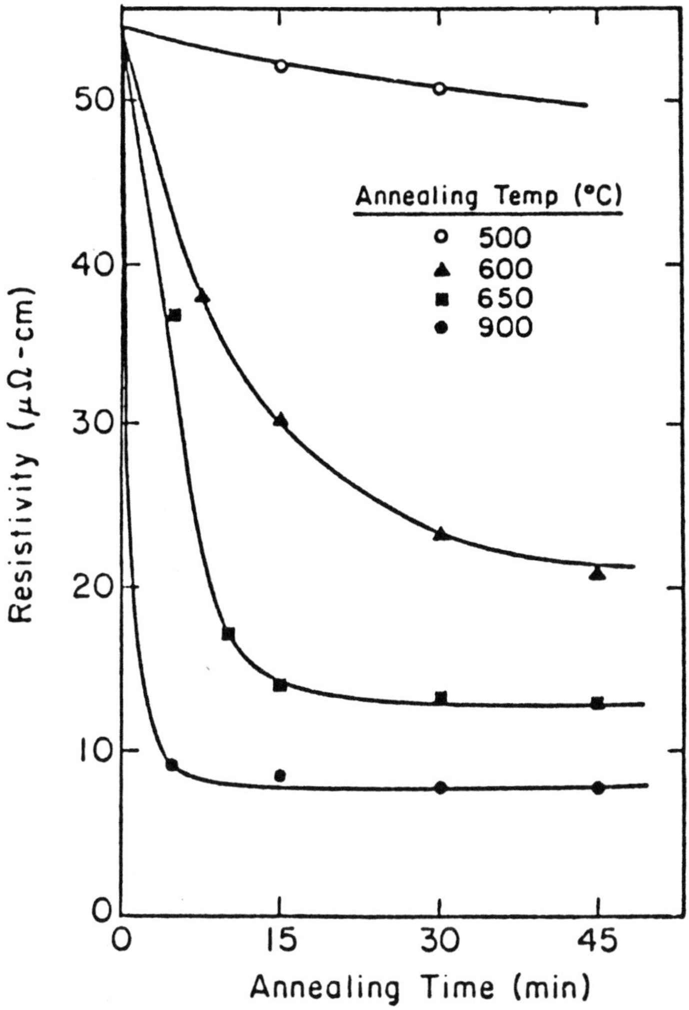

Figure 17: Resistivity of PECVD tungsten versus annealing time and temperature in a 10% H_2/90% N_2 ambient.[16]

reactor. The substrate temperature was kept relatively low, 230°C, the excitation frequency was 13.56 MHz, and pressures were 500 to 700 mTorr.

Deposition rates were found to be on the order of 500 to 600 Å/min, much faster than for pure tungsten. Apparently, adding silicon to the glow discharge enhances deposition rates of tungsten silicide, even for small percentage additions (about 1%). The ratio of W to Si could be adjusted by controlling the amount of SiH_4 added to the WF_6, and the film resistivity depended on this ratio, as shown in Figure 18.

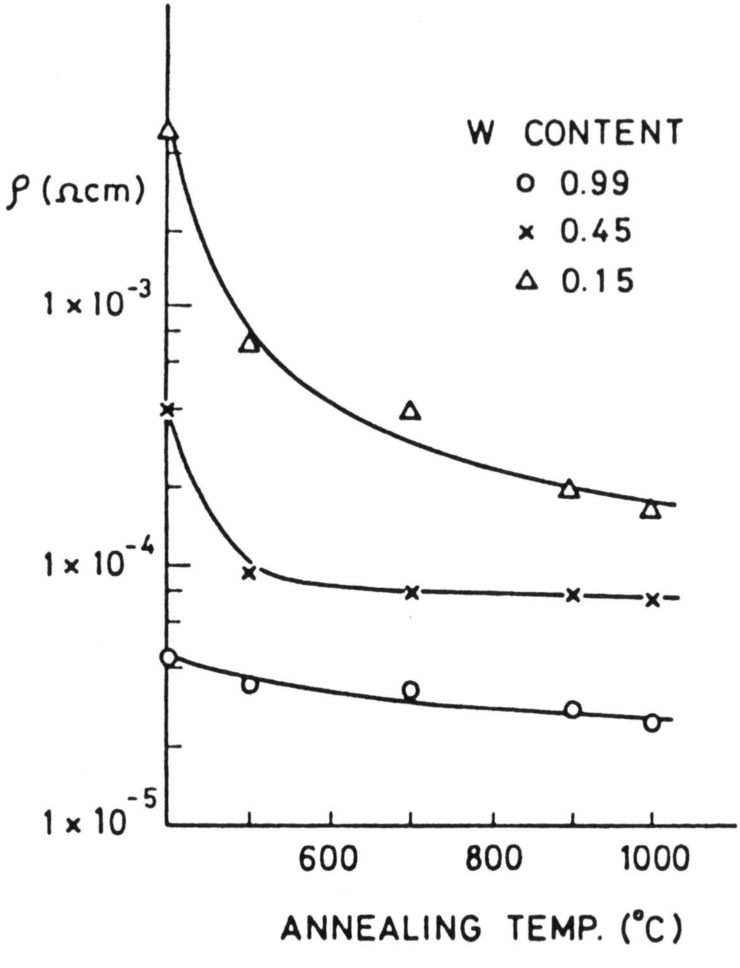

Figure 18: Resistivity versus annealing temperature for different composition WSi_x films annealed in N_2 for 60 minutes.[18]

For films close to stoichiometric, WSi_2, the resistivity even after annealing is ~100 $\mu\Omega$-cm. Since these films were amorphous, both before and after annealing, the reduction in resistivity must have been due to outdiffusion of hydrogen and fluorine. The difference, then, between the observed resistivity (~100 $\mu\Omega$-cm) and the expected value (~50 $\mu\Omega$-cm) may be due to lack of proper crystallization.

5.6.2 Molybdenum

Attempts to deposit PECVD molybdenum have not been successful in producing pure low-resistivity films when MoF_6 + H_2 are used as the reactants.[16]

When the H_2/MoF_6 ratio was below 5:1, the films were porous. Ratios above 7:1 yielded stable films. Analysis indicated a fluorine content in these films as high as 15%. Why such a high concentration of fluorine is found here when tungsten films deposited under similar conditions have less than 1%, is not clear. The measured resistivity of the PECVD molybdenum film (3000 Å) was >10,000 $\mu\Omega$-cm.

A more successful approach to such PECVD of molybdenum films was reported[19] using $MoCl_5$ and H_2. In this case, a sublimator operating at 90° to 100°C was needed to vaporize the solid $MoCl_5$. Again, a parallel-plate, cold-wall plasma reactor was used for the depositions. Temperatures ranged from 170° to 430°C, and the pressure was 1 Torr. The as-deposited films turned out to be amorphous with small amounts of chlorine incorporated. Annealing at high temperatures reduced the resistivity by more than one order of magnitude, as shown in Figure 19, although it is still too high. Also, the resistivity of the as-deposited films decreased as substrate temperature increased. The improved resistivity after heat treatment probably reflects a crystallization of the originally amorphous material, as well as the diffusion of chlorine.

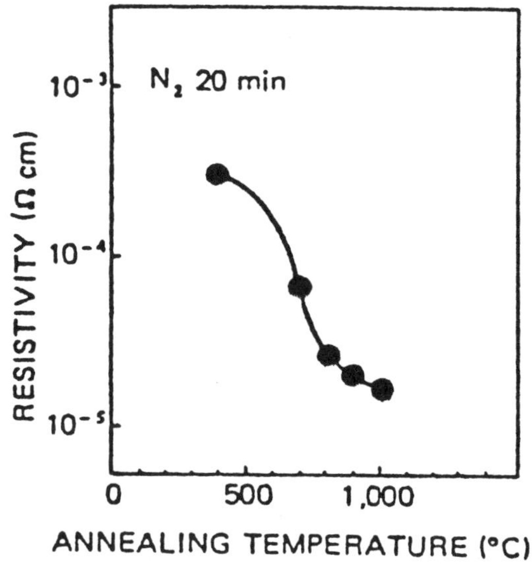

Figure 19: Resistivity of PECVD molybdenum versus annealing temperature.[19]

Molybdenum disilicide was deposited in the same experimental apparatus[19] by adding SiH_4 to the reactant gas mixture. It was possible to vary the $MoSi_x$ stoichiometry by adjusting the flow rates of SiH_4 and H_2. At the same time, the film resistivity also varied, as shown in Figure 20. Unfortunately, the best resistivity achieved was ~800 $\mu\Omega$-cm when a typical thin film value for $MoSi_2$ would be ~100 $\mu\Omega$-cm. The high resistivity may have, again, been due to chlorine trapped in the film. However, it is not clear why the silicide film would differ so radically from the molybdenum film just discussed.

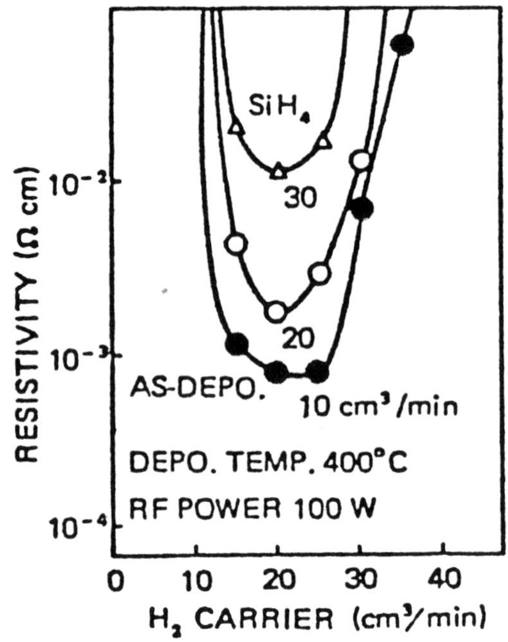

Figure 20: Resistivity of $MoSi_x$ as a function of SiH_4 and H_2 flow rates.[19]

5.6.3 Tantalum

Using plasma enhancement, there is an alternative to the thermal CVD thin film $TaSi_2$ reviewed in the previous chapter.[20] Experiments were carried out in a vertical cold-wall tubular reactor with a single wafer placed normal to the flow. Plasma excitation was provided by a coil wrapped around the quartz tube and positioned above the wafer. The reactive gases flowed from top to bottom. A graphite resistive heater permitted substrate temperatures of 850°C. Excitation frequencies of 600 kHz or 3.5 MHz were used to create the plasma. The source for Ta was $TaCl_5$, and for Si was SiH_2Cl_2. As before, the $TaCl_5$ vapor was produced in a sublimator using H_2 as the carrier gas. The choice of SiH_2Cl_2 rather than SiH_4 was made to minimize gas phase (homogeneous) nucleation. Deposition pressures were in the 1 to 2 Torr range.

These authors found that it was possible to deposit amorphous films whose Ta concentration ranged from 10 to 80 mol % by changing the reactive gas mix. Another feature of the films was that under certain conditions they contained substantial quantities of chlorine and hydrogen. Also, they did *not* adhere to either silicon or silicon dioxide after annealing (argon atmosphere for 1 hour at 900°C). When the substrates were dry etched in an HCl plasma for 2 minutes, they adhered to the substrate even after annealing. Since this etch removed about 50 Å, it appears that the native oxide on the silicon and/or some other surface impurities on both the silicon and silicon dioxide were causing the lack of adhesion.

It was found that the stoichiometry could be adjusted simply by changing the $SiCl_2H_2$ flow rate, as shown in Figure 21. For these curves, the substrate temperature was 580°C, the H_2 flow was 400 sccm, the partial pressure of $TaCl_5$ was 30 mTorr, and the RF power was 10 watts/cm^2 at 3.5 MHz.

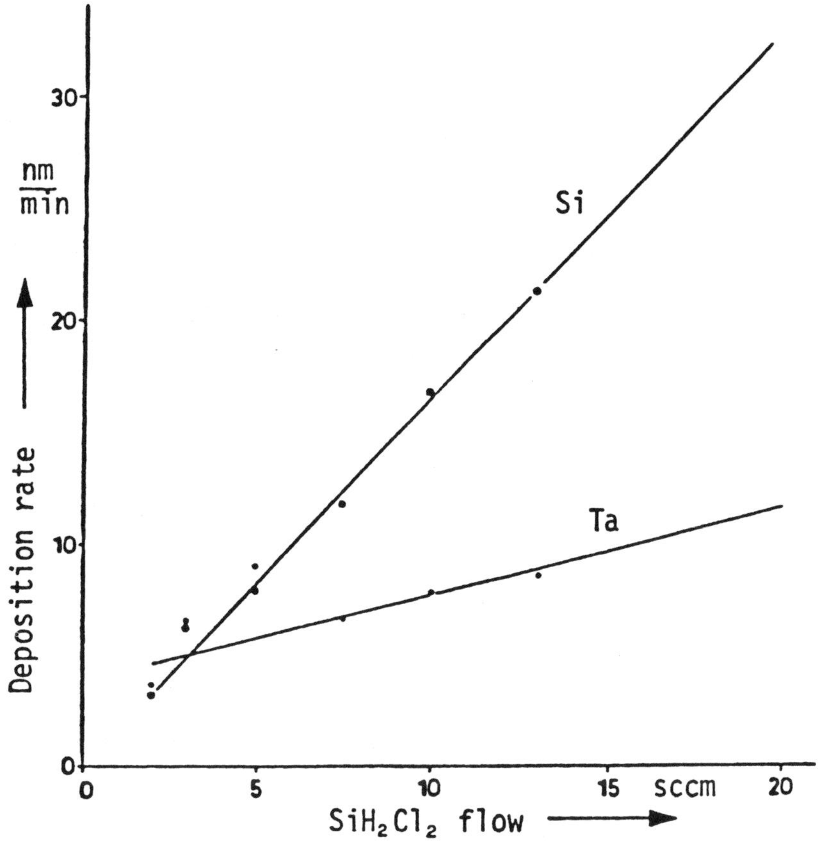

Figure 21: Silicon and tantalum deposition rates versus $SiCl_2H_2$ flow rates.[20]

For the initially amorphous films, annealing converted the film to a polycrystalline one and reduced the resistivity from 180 $\mu\Omega$-cm (as deposited) to 60 $\mu\Omega$-cm. By increasing the substrate temperature to 650°C, the initial films were polycrystalline. By reducing the $TaCl_5$ flow (p_{TaCl_5} ~20 mTorr), increasing the H_2 flow to 800 sccm, and lowering the RF frequency to 600 kHz at 2 watts/cm^2, the authors were able to deposit films with initial resistivities of 70 $\mu\Omega$-cm. Annealing reduced this value to 50 $\mu\Omega$-cm. The interface between the $TaSi_2$ and an underlying layer of polysilicon was smooth, as shown in Figure 22. Also for these conditions no chlorine and only a trace of carbon was detected.

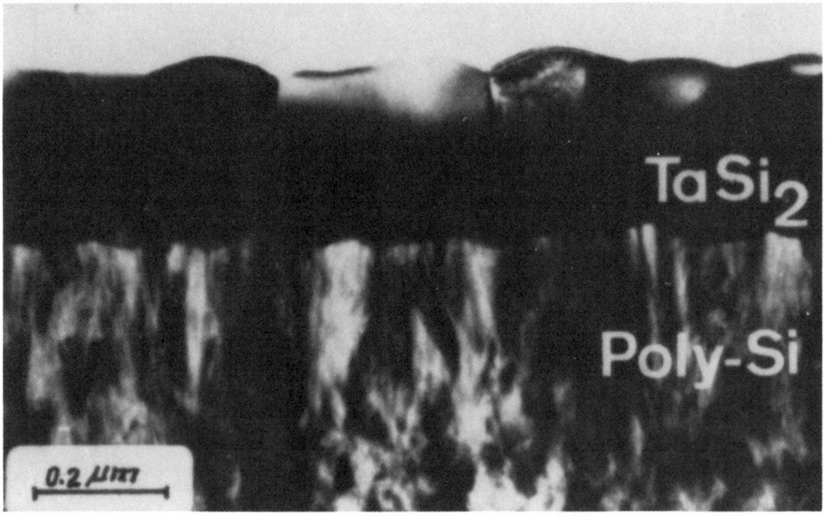

Figure 22: TEM cross section through TaSi polycide structure.[20]

As shown in the previous chapter, similar high-quality $TaSi_2$ polycrystalline films can be prepared by thermal CVD at the same or lower deposition temperature.

5.6.4. Titanium

Recent studies have described the PECVD of titanium disilicide, $TiSi_2$, in a hot-wall tubular reactor.[21,22] This silicide is of particular interest since it has the lowest resistivity of those we have been considering–14 $\mu\Omega$-cm. Although no studies of thermal CVD of $TiSi_2$ films have been reported in the literature, it is claimed[21] that for the $SiH_4/TiCl_4$ reaction, deposition temperatures above 600°C are required, and that this leads to films with larger than desirable grain size. Plasma excitation obviously can lower the deposition temperature, and amorphous $TiSi_2$ films can be deposited. Post deposition anneal can give the desired crystal structure.

Depositions were done at 50 kHz with SiH_4, $TiCl_4$ and Ar as a diluent. At power levels of about 100 watts, the deposition rates were 60 to 80 Å/min. Anneals of the as-deposited films were done at 600° to 650°C. Comparison of the time behavior of resistivity after annealing between $TiSi_2$ and WSi_2 films is shown in Figure 23. Significant is the lower temperature anneal possible for the $TiSi_2$ film and its substantially better resistivity (2:1). As long as the anneal time at 650°C was greater than 5 minutes, it was found that $TiCl_4$ flow (50 to 90 sccm) and deposition temperature (380° to 460°C) had very little effect on the final sheet resistance.

It was found that after annealing the film composition was always stoichiometric, regardless of as-deposited composition, as long as sufficient silicon was available. Some chlorine was detected in the as-deposited films (about

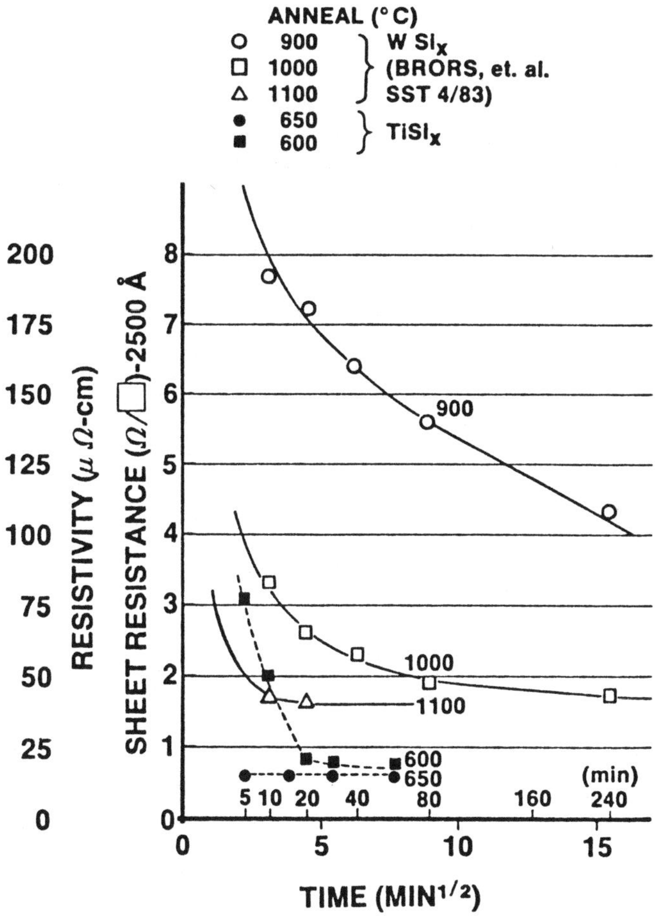

Figure 23: Resistivity versus time for different annealing temperatures and for WSi_x and $TiSi_x$.[21]

1%), and this value was reduced well below this level after annealing.[22] The as-deposited film was really a sandwich of three layers. The top layer was thin amorphous silicon covered by several Å of silicon dioxide, the intermediate one was an amorphous silicide layer of composition $Ti_{0.94}Si$, and the bottom one was another layer of amorphous silicon of equal thickness. After annealing, the average grain size on a polysilicon substrate was 5000 Å, and on a single-crystal substrate was 1000 Å. Also, dopants from the substrates did not diffuse up to the silicide layers during annealing.

Oxidation behavior of the $TiSi_2$ on silicon is the final concern.[22] A good deal of surface and interfacial roughness was observed on the films that had been put down on silicon and wet oxidized for one hour at 800°C. The oxide layer was about 1800 Å thick. Further study of the quality of the oxide layer will be necessary before this film can be considered for device applications.

5.7 ALUMINUM

Again, in the previous chapter we saw that deposition of thermal CVD aluminum films is possible. Clearly, they improve the step coverage problem, since they are reasonably conformal. However, they are not completely satisfactory because of the inability to readily dope them with copper to limit electromigration, and they exhibit grain sizes that are too large.

As we saw in the $TiSi_2$ case just covered, it may be possible to deposit an amorphous film and anneal, if necessary, to achieve a desirable grain size. In fact, PECVD of aluminum films composed of small crystals with random orientation have been obtained[23] with as-deposited resistivities of 5 to 10 $\mu\Omega$-cm.

Depositions were obtained from either $AlCl_3$ or $Al(CH_3)_3$ mixed with either Ar or H_2, or both. The plasma was excited by a 13.56 MHz source. At a power level of 200 watts, the deposition rate was 250 Å/min. Step coverage over 1 micron high steps was at least 50%.

No attempts were made to dope these films (electromigration), but since one of the key features of PECVD is the ability to control film composition, it may be possible to do this.

REFERENCES

1. Rosler, R.S., Benzing, W.C. and Baldo, J., A production reactor for low temperature plasma enhanced silicon nitride deposition. *Solid State Technol.* 19(6):45 (1976).
2. Hollahan, J.R., Wauk, M.T. and Rosler, R.S., Plasma-enhanced chemical vapor deposition of thin films and some of their etching characteristics, in *Proceedings of the Sixth International Conference on Chemical Vapor Deposition,* eds. Donaghey, L.F., Rai-Choudhury and R.N. Tauber, Electrochem. Soc., Pennington, NJ, 1977, p. 224. This table was originally presented at the Fall 1977 Meeting of The Electrochemical Society, Inc. held in Atlanta, Georgia.
3. Claasen, W.A.P., Valkenburg, W.G.J.N., Willemsen, M.F.C., and v.d. Wijgert, W.M., Influence of deposition temperature, gas pressure, gas phase composition, and RF frequency on composition and mechanical stress of plasma silicon nitride layer. *J. Electrochem. Soc.* 132:893 (1985).
4. Claasen, W.A.P., Valkenburg, W.G.J.N., Habraken, F.H.P.M., and Tamminga, Y., Characterization of plasma silicon nitride layers. *J. Electrochem. Soc.* 130:2419 (1983).
5. Chow, R., Lanford, W.A., Ke-Ming, W. and Rosler, R.S., Hydrogen content of a variety of plasma-deposited silicon nitrides. *J. Appl. Phys.* 53:5630 (1982).
6. Fang, Y.K., Huang, C.F., Chang, C.W. and Lee, R.H., Preparation and characterization of boron- and phosphorus-doped hydrogenated amorphous silicon nitride films. *J. Electrochem. Soc.* 132:1222 (1985).
7. Allaert, K., Van Calster, A., Loos, H. and Lequesne, A., A comparison between silicon nitride films made by PECVD of N_2-SiH_4/Ar and N_2-SiH_4/He. *J. Electrochem. Soc.* 132:1763 (1985).

8. van de Ven, E.P.G.T., Plasma deposition of silicon dioxide and silicon nitride films. *Solid State Technol.* 24(4):167 (1981).
9. Adams, A.C., Alexander, F.B., Capio, C.D. and Smith, T.E., Characterization of plasma-deposited silicon dioxide. *J. Electrochem. Soc.* 128: 1545 (1981).
10. Yokoyama, S., Dong, D.W., DiMaria, D.J. and Lai, S.K., Characterization of plasma-enhanced chemically-vapor-deposited silicon rich silicon dioxide/thermal silicon dioxide dual dielectric systems. *J. Appl. Phys.* 54:7058 (1983).
11. Avigal, I., Inter-metal dielectric and passivation-related properties of plasma BPSG. *Solid State Technol.* 26(10):217 (1983).
12. Tong, J.E., Schertenleib, K. and Carpio, R.A., Process and film characterization of PECVD borophosphosilicate films for VLSI applications. *Solid State Technol.* 27(1):181 (1984).
13. Nguyen, V.S., Burton, S. and Pan, P., The variation of physical properties of plasma-deposited silicon nitride and oxynitride with their compositions. *J. Electrochem. Soc.* 131:2348 (1984).
14. Reif, R., Plasma enhanced chemical vapor deposition of thin crystalline semiconductor and conductor films. *J. Vac. Sci. Technol.* A2(2):429 (1984).
15. Donohue, T.J., Burger, W.R., and Reif, R., Low-temperature silicon epitaxy using low pressure chemical vapor deposition with and without plasma enhancement. *Appl. Phys. Lett.* 44:346 (1984).
16. Hess, D.W., Plasma-enhanced chemical vapor deposition of metal and metal silicide films, in *Proceedings of the Matl. Res. Society Symposium,* 1985, Vol. 38.
17. Tang, C.C., and Hess, D.W., Plasma Enhanced chemical vapor deposition of β-tungsten, a metastable phase. *Appl. Phys. Lett.* 45(6):633 (1984).
18. Akitomoto, K., and Watanabe, K., Formation of W_xSi_{1-x} by plasma chemical vapor deposition. *Appl. Phys. Lett.* 39:445 (1981).
19. Tabuchi, A., Inoue, S., Maeda, M. and Takagi, M., Formation of Mo and Mo-silicide film by plasma assisted chemical vapor deposition, in *Proceedings of 23rd Symposium on Semiconductor and IC Technology of Japan,* 1982, p. 60.
20. Hieber, K., Stolz, M. and Wieczorek, C., Plasma enhanced chemical vapor deposition of $TaSi_2$, in *Proceedings of Ninth International Conference on Chemical Vapor Deposition,* eds. Robinson, McD., Cullen, G.W. (The Electrochemical Society, Pennington, NJ, 1984), Vol. 84-6, p. 205. This figure was originally presented at the Spring 1984 Meeting of The Electrochemical Society, Inc. held in Cincinnati, Ohio.
21. Rosler, R.S. and Engle, G.M., Plasma enhanced CVD of titanium silicide. *J. Vac. Sci. and Technol.* B2(4):733 (1984).
22. Morgan, A.E., Stacy, W.T., DeBlasi, J.M., and Chen, T-Y. J., Material characterization of plasma-enhanced chemical vapor deposited titanium silicide. *J. Vac. Sci. and Technol.* B4(3):723 (1986).
23. Ito, T., Sugii, T. and Nakamura, T., Aluminum plasma-CVD for VLSI circuit interconnections, in *Digest of Papers of 1982 Symposium on VLSI Technology* (IEEE, NY, 1982).

6

Production CVD Reactor Systems

6.1 INTRODUCTION

In Chapters 1 and 2, we not only covered the basics of thermal and plasma-enhanced CVD, but we described the general reactor configurations that researchers have explored over the years. From these concepts have come a few production CVD reactors that satisfy the commercial needs of the integrated circuit manufacturing process.

When considering a production reactor, we first assume that the requisite quality film can be made at least one at a time. The challenge then is to develop a reactor that is capable of acceptable wafer throughput with each wafer having film thickness within an acceptable tolerance. For example, we may want a reactor that can process 30 wafers per hour with thickness uniformity on a single wafer, and from wafer to wafer, of ± 5%. In addition, we may impose other conditions such as permissible number of particles per cm^2, or for epi silicon films, the allowable number of defects per cm^2. When we speak of wafer throughput, we are concerned with the actual cost per wafer for this process step.

The cost per wafer will depend on many factors. First, the reactor can be quite expensive, so it is a capital item and must be amortized. Also, if the reactor has to be cleaned very frequently or is unreliable and experiences a lot of down time, then this will also add to the capital cost. If the reactants are expensive and not utilized efficiently, then this is another expense item. Energy requirements can be high for heating either the chamber or the susceptor. So, a system with high wafer throughput leads in the direction of lower cost per wafer, provided film quality is acceptable.

One way to achieve high wafer throughput is to pack many wafers into each reactor load. This is what is referred to as a "batch" system. It is for this reason that most production CVD reactors today operate as batch systems. One alternative is to operate a single-wafer system, but do it in a continuous fashion; two such systems will be described. Finally, if the process is sufficiently sensitive, it may be acceptable to operate a single-wafer system without the continuous operation, and several newer systems will be reviewed later.

It is important to recognize that a production reactor is not simply a reaction chamber. If it is a low pressure unit, there will be a vacuum system which can be quite complex. There will be a gas panel which regulates gas mass flow into the chamber. The method of heating the wafers and/or the entire chamber has to be chosen carefully. Wafer transport involves many tradeoffs, and for batch systems if any degree of automation is required, will be quite involved. Finally, most production reactors these days operate under microprocessor control, and quite a lot of software must be developed.

In the balance of this chapter, we will review a number of reactors that are currently in production use. As much as possible, the above features of such reactors will be discussed.

6.2 LOW-TEMPERATURE SILICON DIOXIDE REACTORS

One of the earliest production-ready CVD reactors was developed to deposit low-temperature SiO_2 at atmospheric pressure.[1] Rather than a batch system, Applied Materials, Inc. designed a system with a large area of uniform reactant flow and moved wafers continuously through this zone. The wafer remained in this reactant flow long enough to achieve the required film thickness, typically 1 micron. A block diagram of this reactor is shown in Figure 1, showing the major subsystems.

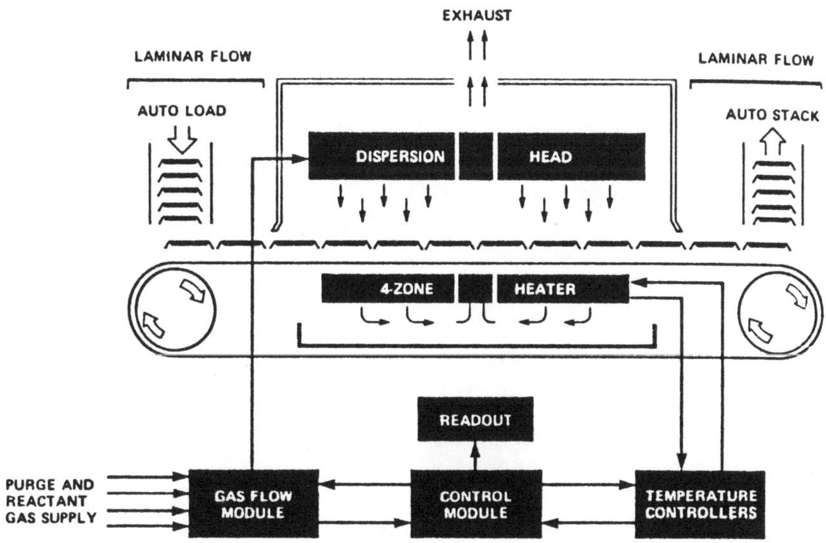

Figure 1: Continuous atmospheric pressure CVD reactor.[1]

They are:

(1) Reactant/purge gas disperser

(2) Heater/temperature control

(3) Wafer transport

(4) Wafer tray load/unload

(5) Reactant/purge gas flow control

(6) Reactor exhaust

Wafers are transported, in Inconel trays, on a moving belt through a reactant gas flow while being heated from below by quartz radiant heaters. Reactant gases are introduced through a unique disperser head design. Details of the gas introduction are shown in Figure 2.

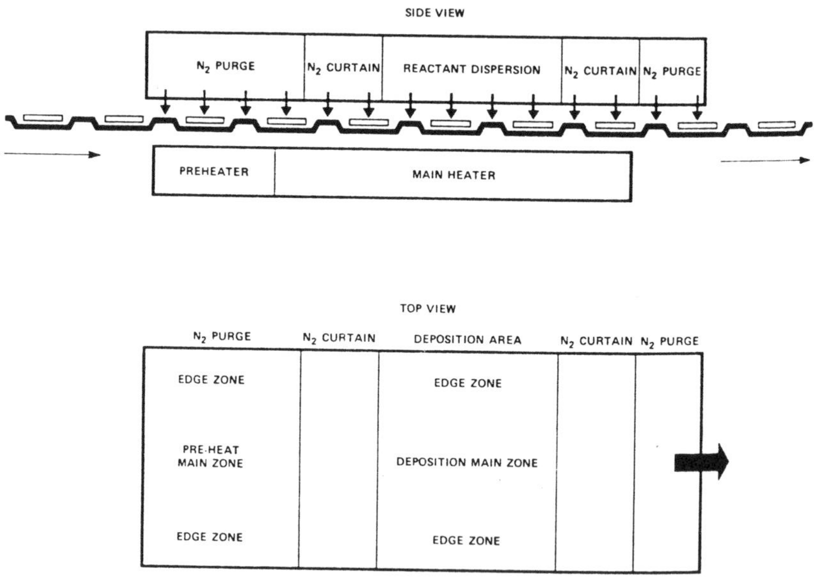

Figure 2: Gas flows in AMS-2000.[1]

As the wafer tray (which can hold up to three wafers) enters the reactor, it passes through a nitrogen purge and curtain, so that the wafers heat up in an inert atmosphere before being exposed to the reactant gases. It also prevents air from entering the reaction zone. After deposition, the wafer tray exits through a similar nitrogen and purge. The flow and temperature profiles along the wafer path are shown in Figure 3.

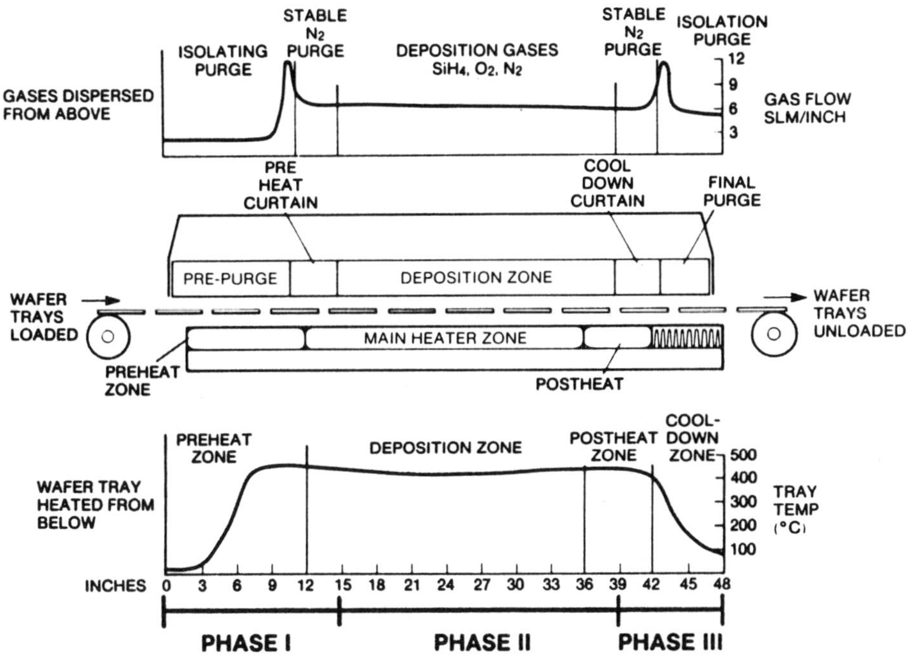

Figure 3: Gas flow diagram—Applied Materials.

The reactant gas disperser is assembled from anodized aluminum plates of alternating geometry. A cross sectional view normal to the flow direction illustrating the gas flow pattern is shown in Figure 4.

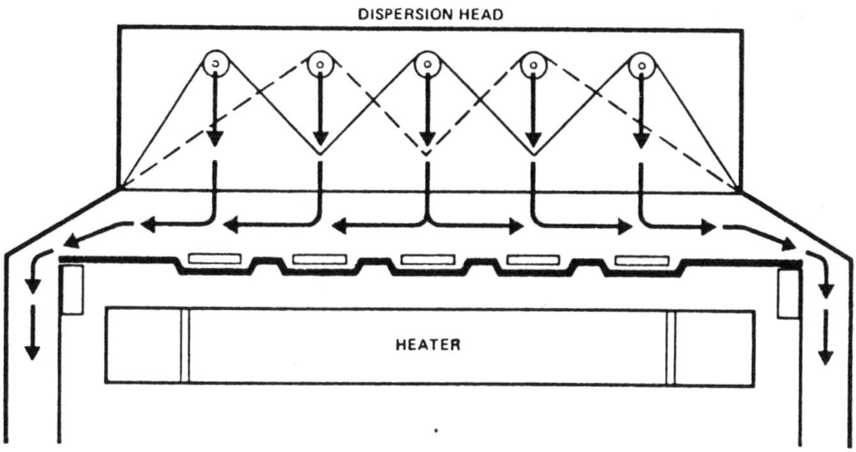

Figure 4: AMS-2100 gas flow pattern.[1]

The gases are introduced via five perforated tubes positioned in the flow direction, so that a uniform flow can be established. A short flow path to the wafer surface is chosen to minimize any gas phase reactions. The reactant gases are rapidly removed from the reaction zone and exhausted downward.

A photograph illustrating the deposition zone of an AMS-2100 is shown in Figure 5. Here, the N_2 purge regions and the exit from the dispersion head can be clearly seen.

Figure 5: Deposition region of AMS-2100. Applied Materials.

Approximately once each shift, the disperser head has to be cleaned. A vacuum cleaner is used to remove SiO_2 particles that have built up on the disperser lower surface.

This basic system was designed to deposit SiO_2 from the $SiH_4 + O_2$ reaction at about 400°C and atmospheric pressure. It can also deposit doped oxides by introducing PH_3 for phosphorus doping or B_2H_6 for boron doping. In order to protect personnel from these toxic dopants, the reactor is housed in a vented enclosure.

Due to the high deposition rates possible at atmospheric pressure, approximately 1000 Å/min, wafer throughput can be as high as 200 to 400 per hour. Also, since this is an atmospheric pressure reactor, there is no expensive vacuum system, and the capital cost of the reactor system is modest. These two facts contribute to a low cost per wafer processed, and has allowed this system to remain in commercial use for over 13 years.

The shortcomings of such a system are that some of the SiO_2 particles formed on the disperser head wind up on the wafers. As IC dimensions have shrunk over the years, the ability to tolerate even very small particles has lessened, and other approaches have been developed.

A second atmospheric pressure system has seen some considerable commercial success in recent years. It is based on the concept of mixing the reactants at the substrate surface, following the concept illustrated in Figure 25 of Chapter 1.[2] As before, the wafer travels through the region of reactant gas flow on a belt, passing through nitrogen curtains. A sketch of the CVD chamber of this reactor is shown in Figure 6. As we see, the reactive gases enter through three feed lines, and they are not mixed until they exit the ejector. As soon as these gases impinge on the wafer, they are exhausted to the vent.

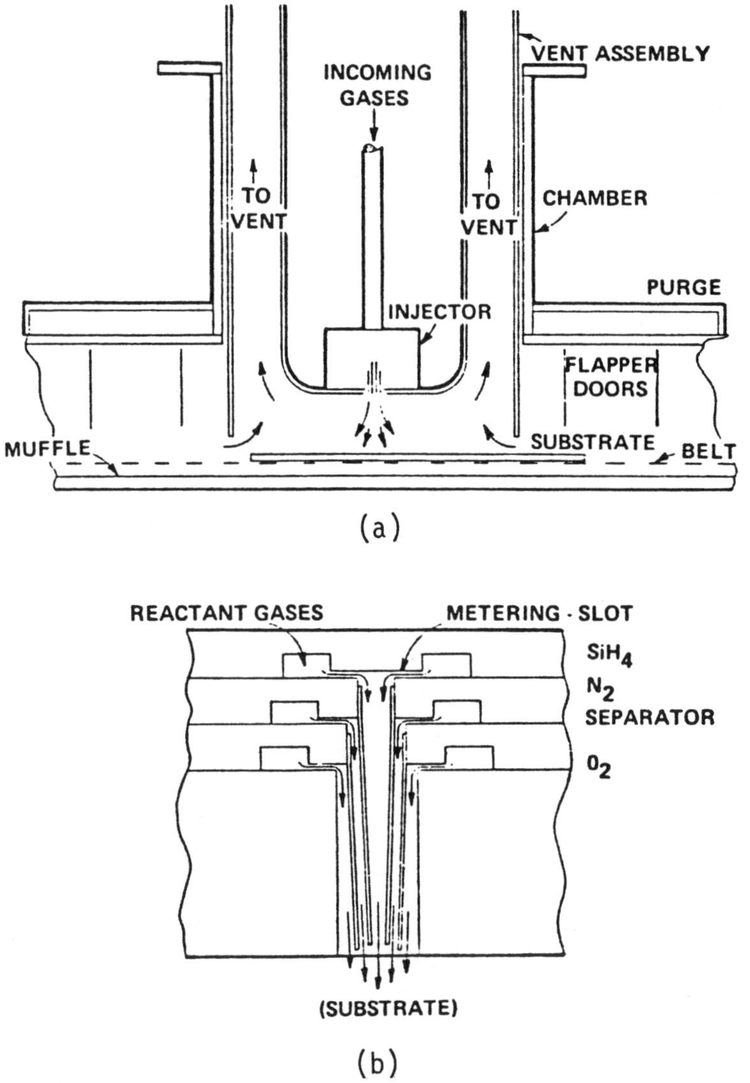

Figure 6: Atmospheric pressure CVD system with gas injector.[2]

Pressure in the purge plenums is arranged to be slightly higher than pressure at the exit to the injector. In this way, none of the reactive gases can flow anywhere but the vent.

By minimizing the residence time of the reactive or spent gases above the wafer, the designers of this reactor attempt to minimize gas phase reactions and subsequent generation of particles. Since the SiH_4 and O_2 are introduced unmixed just above the wafer at high velocity, deposition rates as high as 10,000

Å/min are achieved. This permits a throughput of 96-4" wafers per hour with a 6" belt.

Since the chemistry is very similar to the previous system, SiO_2, PSG or BPSG films can be deposited at temperatures of approximately 350° to 450°C. It is very interesting that films deposited by this technique exhibited lower tensile stresses than other techniques, about 1 x 10^{-9} dynes/cm^2.

The primary advantage of this reactor over the Applied Materials System is that it requires less power, and the dispersion head is easily removed for cleaning. However, the AMS-2100 has about half the footprint, so it takes up less valuable clean room space. As far as film quality is concerned, both systems produce comparable films and particles remain a concern in each system.

6.3 HOT TUBE, LOW PRESSURE, THERMAL SYSTEMS

In Chapter 2, we reviewed the concept of carrying out CVD processes at low pressure so that deposition becomes surface controlled. When the only thing controlling the uniformity of deposition is the temperature of the wafer surface, all we have to do is ensure that the wafer is in a uniform temperature furnace. Again, at low pressures, the diffusion coefficient is so large that we can stack wafers up next to each other so 50 to 100 can be placed in a long tubular furnace.

Because of this arrangement, it has become economically desirable to deposit poly, SiO_2 and Si_3N_4 in such furnaces, and there are many commercial versions of this type of furnace. A typical LPCVD furnace is shown in Figure 7.

Figure 7: Tylan-Tytan automated LPCVD reactor.

This system contains a four-tube stack of furnace tubes for even higher productivity. Wafers are stacked in a quartz boat which can be loaded from left to right into the furnace tube. In any CVD system, deposition occurs on the inside of the tube, and it is not economical to clean the tube after each deposition. The film that forms on the tube can be scraped as the boat is inserted, and this leads to particles which end up on the wafers. To avoid this, the wafer boat is loaded by a cantilever arrangement such as one shown in Figure 8. In this figure, we also see how each furnace tube is constructed.

An electrical resistance heater with more turns at the tube ends (to compensate for heat losses) surrounds each tube. There is a vertical laminar flow hood over the loading area to minimize particle contamination of the wafers being loaded. As we can see, there are temperature controls for the furnace tubes, and a power module to provide the electrical power. When operated as a LPCVD system, a unit including both the gas flow and vacuum systems is positioned on the right side. Such a unit is shown in Figure 8. Here we can see the vacuum pumps on the left, and the mass flow controllers on the right. The vacuum pump oil recirculation systems are shown in the slide out drawers. As can be seen in Figure 9, this system, as well as most current similar systems, operate under computer control.

As before, the major design challenge in such LPCVD systems is to maintain uniform film quality and deposition rate on all wafers in the system. Since the reactant gas flow can deplete as it travels down the tube, it is not always easy to achieve this goal with tight tolerances. In some cases, changing the temperature along the tube can compensate for an axial change in deposition rate.

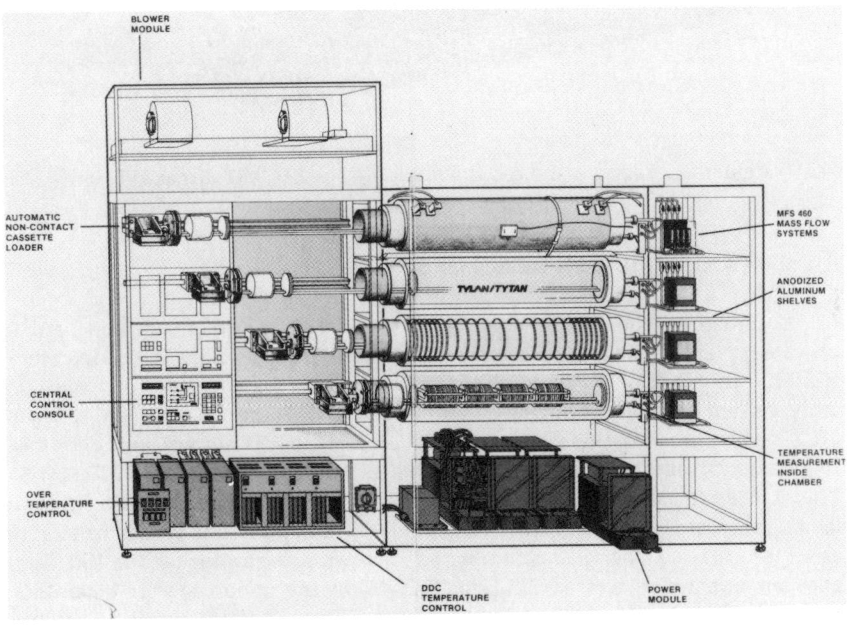

Figure 8: Tylan-Tytan LPCVD reactor details.

Figure 9: Vacuum and gas flow module for LPCVD furnace, Tylan.

6.4 EPITAXIAL SILICON REACTORS

Commercial epitaxial silicon reactors are difficult to design and hard to operate reliably because of the very high temperatures they must operate at (~1000° to 1200°C). One of the more successful systems has been the radiantly heated reactor developed by Applied Materials, Inc. and shown in Figure 10. The system has the same elements as the LPCVD system just described. A schematic showing how these elements are arranged is shown in Figure 11. This is a cold-wall barrel reactor system with a 3-zone bank of high intensity lamps to provide the heat to the wafers on the susceptor. The system can be run as either an atmospheric pressure or reduced pressure reactor (about 100 Torr), and can operate with either $SiCl_4$ or $SiCl_2H_2$ as the silicon source. When $SiCl_4$ is used, a bubbler to vaporize it must be provided. A more detailed sketch of the reactor module is shown in Figure 12. The susceptor is hung vertically and

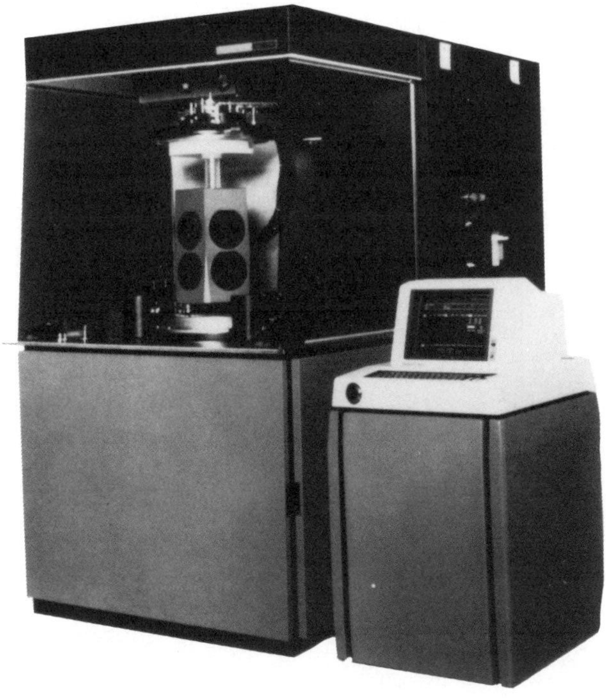

Figure 10: AMC 7800 radiantly-heated epitaxial reactor system—Applied Materials, Inc.

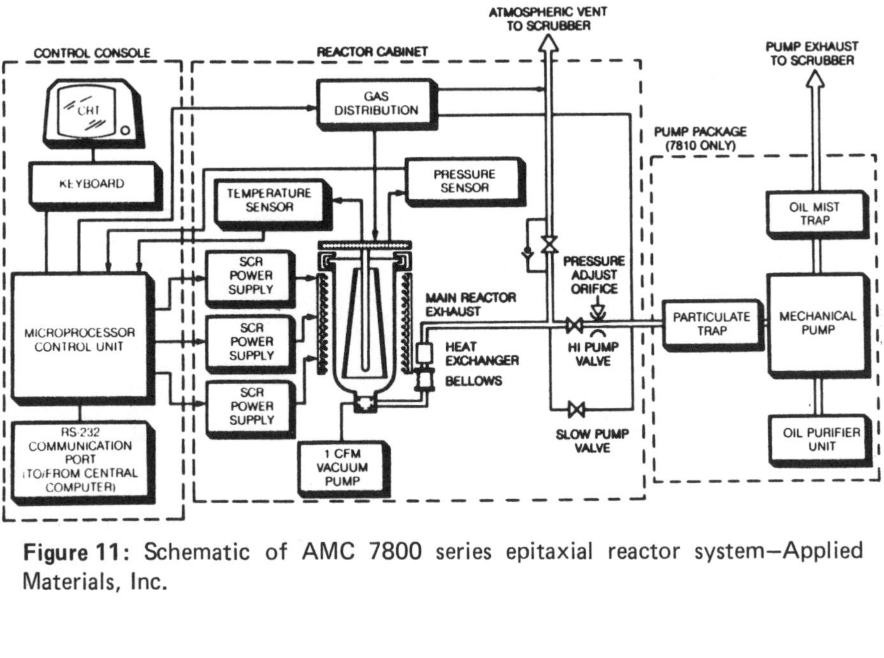

Figure 11: Schematic of AMC 7800 series epitaxial reactor system—Applied Materials, Inc.

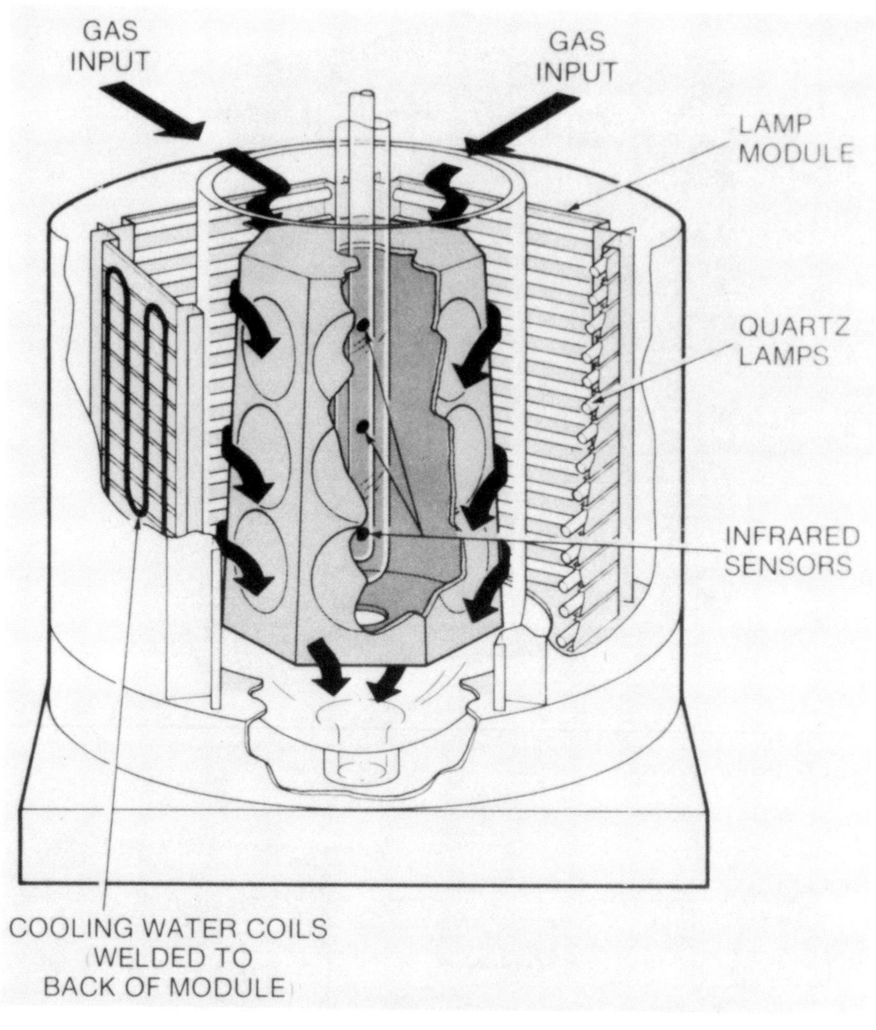

Figure 12: Reaction chamber of AMC 7800 epi reactor system—Applied Materials, Inc.

rotated during processing. Heat is provided by radiant heating modules whose radiation passes through the quartz bell jar and is absorbed by the silicon carbide-coated graphite susceptor. Air flow is used to cool the bell jar to minimize silicon deposition there. The radiant heating modules are gold coated to reflect thermal radiation returning from the hot susceptor. This is a workable arrangement, as this gold coating is transparent to the high-intensity optical radiation of the lamp assembly, but reflects the thermal radiation of the susceptor.

The effect of heating this axisymmetric arrangement from the outside is to create a very uniform temperature environment for each wafer. By this means, thermally-induced slip is virtually eliminated.

At the pressures at which this system is run, deposition is diffusion controlled. Therefore, the flow patterns set up and boundary layer thicknesses are important to the goal of uniform deposition on all wafers. As noted in Chapter 3, the narrowing flow passage in the flow direction compensates for reactant depletion.

As wafers get larger, fewer wafers per load can be handled. For example, the susceptor can have 10 faces with 4 wafers per face for 3" wafers, for a total of 40 wafers per load. For 5" wafers, the susceptor will have 5 faces and 2 wafers per face, for a total of 10. This loss of capacity can only be compensated for by building a larger reactor or by purchasing two systems.

The other configuration of silicon epi reactor that has been successful commercially is the so-called pancake reactor. The basic concept behind the system was shown in Chapter 1, Figure 22. As a production reactor, it is simple, rugged and inexpensive compared to the radiantly-heated barrel just described. The typical graphite wafer holder inside a quartz bell jar is shown in Figure 13.

Figure 13: Vertical pancake epi silicon reactor chamber—Gemini Research.

These systems have been the work horses for the less demanding epi applications. In general, epi thickness uniformity has not been as good as the radiative systems, but the biggest difficulty has been that they often produce wafers with significant slip defects. As noted earlier in Chapter 3, recent design modifications have tended to minimize these problems.

A new epi reactor, the "Precision Epi 7010," has recently been introduced by Applied Materials. The unusual layout of this new system is shown in Figure 14. The central concept in this new reactor system is lower processing cost per wafer, while hopefully maintaining the same quality of epi films. The design features that lower the per wafer cost are the dual susceptor design and the use of low-frequency induction for heating. With two susceptors, one can be loaded or unloaded while the other is being processed.

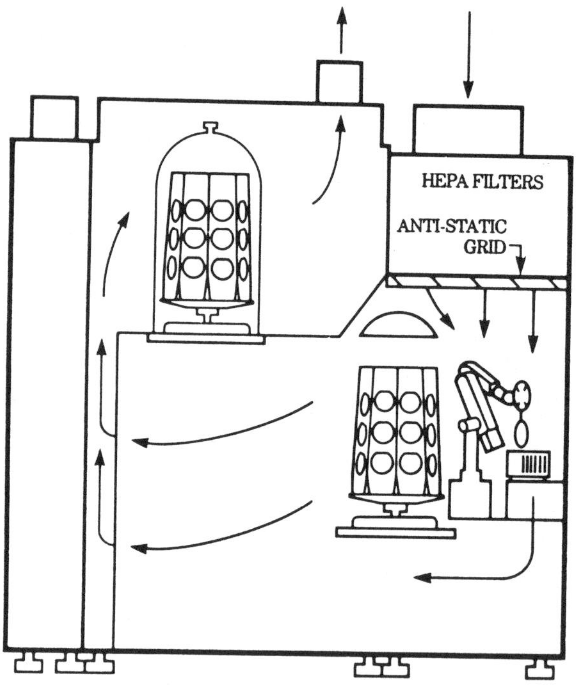

Figure 14: Schematic of Applied Materials Precision Epi 7010 reactor system.

The high-intensity lamp array in the earlier design is expensive, so they have been replaced by a 10 kHz RF induction coil placed around the entire bell jar. The bell jar and coil are shown in Figure 15. By using this arrangement and an electrically-conducting silicon carbide-coated graphite susceptor, the coil heats the susceptor by inducing currents to flow in it. The bell jar is then gold coated to reflect thermal radiation back onto the wafers for a more uniform thermal environment than is usually achieved with induction heating. Water cooling on the outside of the bell jar is used to minimize deposits and subsequent cleaning operations. This concept of dual direction heating is illustrated in Figure 16.

Figure 15: Applied Materials Precision Epi 7010 bell jar and induction coil.

In addition to the design goal of reducing the processing cost per wafer, this system has been developed to reduce the particulate contamination to a minimum. To accomplish this, the entire system is enclosed within a laminar flow hood, so that the wafers are handled in a class 10 environment. Secondly, wafers are loaded and unloaded by an automated robot, so that human intervention is not necessary to handle the two susceptors. Both of these features are illustrated in Figure 16.

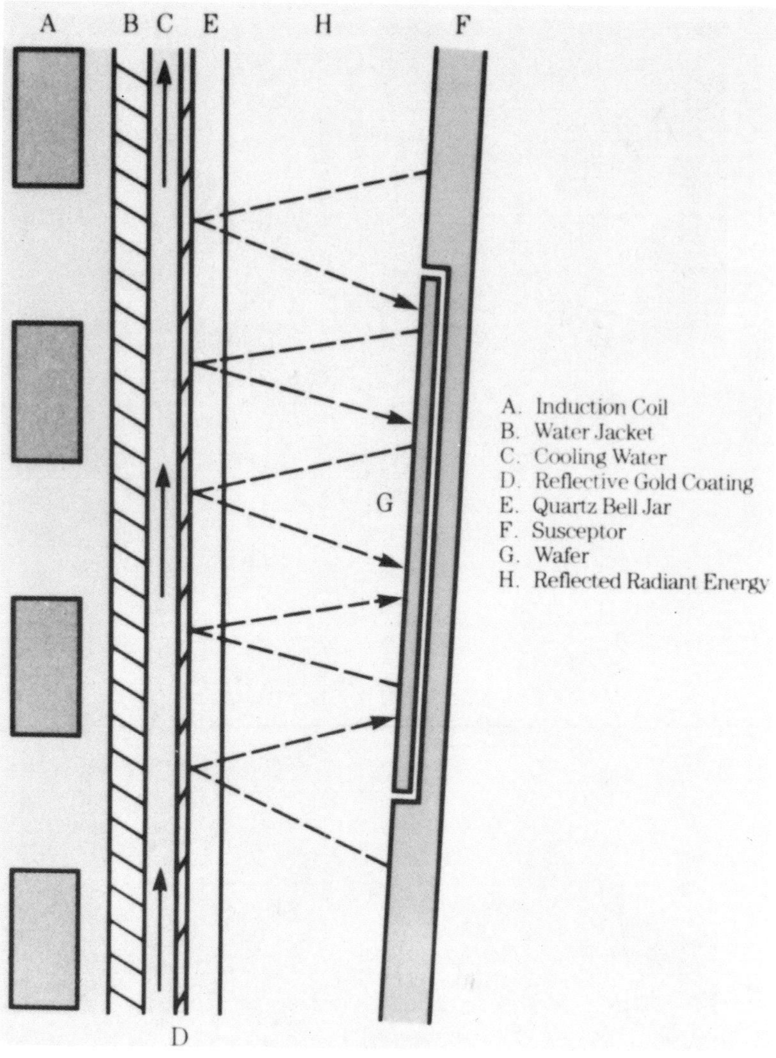

Figure 16: Dual-direction heating in Applied Materials Epi 7010 reactor system.

Finally, clean room floor space is expensive, so this system has been designed so that the entire reactor can be outside the clean room, with one face occupying wall space only.

Another new epi reactor has also been introduced recently by Gemini Research. This system uses a cold quartz bell jar much the same as the earlier pancake reactor. The susceptor configuration is new, however, and is shown in Figure 17.

Figure 17: Reactor chamber of Gemini-Tetron I epi system.

Just as in the new Applied Materials reactor, the problem being addressed is the reduced throughput with conventional epi reactors as wafers get larger. In this reactor configuration, 50 wafers can be placed on the 25 tapered cavities placed radially within the bell jar. There are resistance heaters above and below the silicon carbide-coated graphite wafer holders, and heat loss at the outer periphery is compensated for with external heat lamps.

Other features of this system are comparable to the Applied Materials Epi 7010, in that robotic wafer loading is included, and the system is fully computerized as well as designed for through-the-clean-room-wall operation.

6.5 PLASMA-ENHANCED SYSTEMS

There have been a wide variety of plasma-enhanced CVD reactors developed over the years, but only a few have had significant success as production systems.[3] The first commercially successful plasma-enhanced reactor system was a parallel-plate, capacitively-coupled, cold-wall system based on the concept described in Chapter 2, Figure 13. A block diagram of the complete system is shown in Figure 18. In this application, a 50-kHz RF power supply is used to power the upper electrode, while the lower electrode and the balance of the chamber is grounded. For this system, it was felt that it was necessary to rotate the lower electrode to smooth out the nonuniformities. As a result, an elaborate magnetically-coupled rotation mechanism is used (to preserve vacuum integrity) to accomplish this. The major disadvantage of such elec-

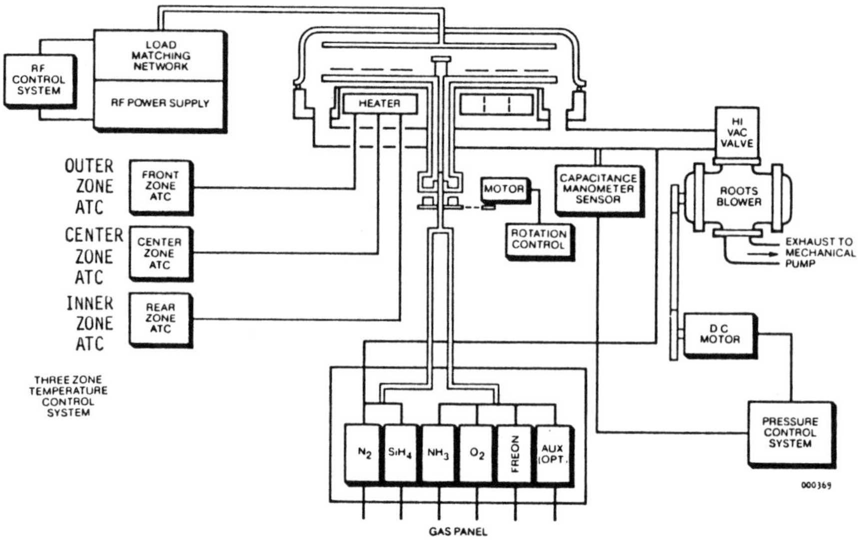

Figure 18: Applied Materials AMP-3300 plasma CVD system.

trode rotation is that it is inconvenient to position the electrode heater on the underside of the electrode. Therefore, the three-zone heater heats the electrode by radiation, being about ¼-inch from it.

Gas enters at the center of the electrode along the axis of rotation from the gas panel. It was found that if the gas were allowed to flow in directly and impinge normally to the upper electrode, that excessive deposits built up rapidly at the center of this upper electrode, and poor deposition uniformities were observed. Therefore, a gas injection shield was placed at the center of the lower electrode to provide some radial momentum to the incoming gas flow. This shield is illustrated in Figure 19.

The system operates at about 400 mTorr and with flow rates of about 1 slm, so a roots blower is used to pump the system to base pressure (~1 mTorr) and then handle the gas flow at these low pressures.

The top of the reaction chamber is hinged to permit access to the lower electrode (platen) that holds the wafers. This arrangement is illustrated in Figure 20. The platen is anodized, while the upper electrode is bare aluminum. The upper electrode housing is flame spray coated with aluminum oxide to eliminate electrical discharges above the upper electrode. The platen has pockets to accept wafers and maintain them in fixed positions as it rotates.

As noted earlier, wafer throughput is a key concern. The more wafers that can be placed on the platen, the higher the throughput. For 3" wafers, a load of 42 wafers could be handled. As wafer sizes have increased, however, the capacity of this system has been severely impacted. For 5" wafers, only 16 can be loaded at one time. Obviously, the problem is more difficult for 6" and eventually 8" wafers. This difficulty eventually led to the development of the hot tube PECVD system to be discussed next.

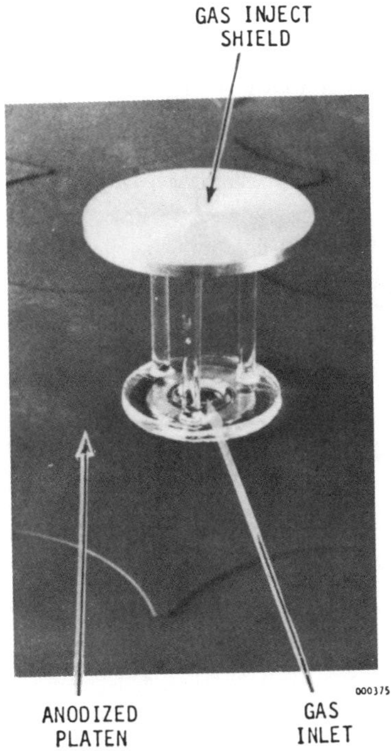

Figure 19: AMP-3300 gas inject shield. Applied Materials, Inc.

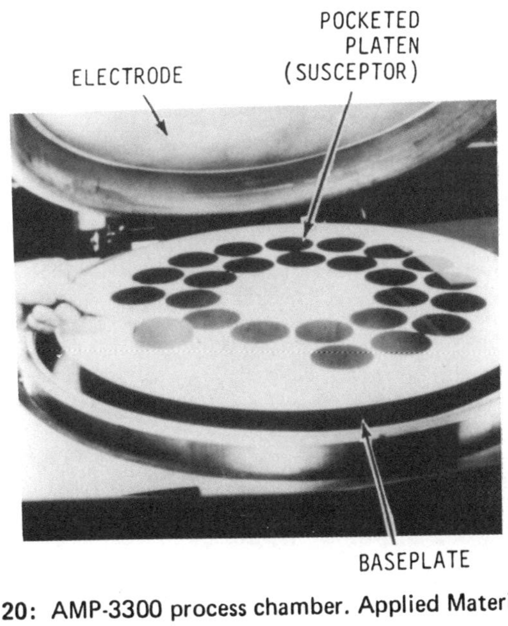

Figure 20: AMP-3300 process chamber. Applied Materials, Inc.

It should also be noted that this reactor has to occupy space in the clean room, and is not operated under computer control. For application to modern fab lines, these are two disadvantages of this system.

The hot wall approach to the plasma-enhanced CVD system has been described in Chapter 3. A schematic of a typical system is shown in Figure 21. The elements of this system are similar to that of the cold-wall system just described. There is a gas panel, vacuum system, and an RF power supply to create the discharge. The RF frequency typically used is 400 kHz. The reaction chamber of such a system is shown in Figure 22. The electrodes are a set of several long narrow rectangular slabs of graphite with pockets cut into them. The graphite electrodes lead to some problems with particulate contamination, but attempts to use aluminum have not been successful.

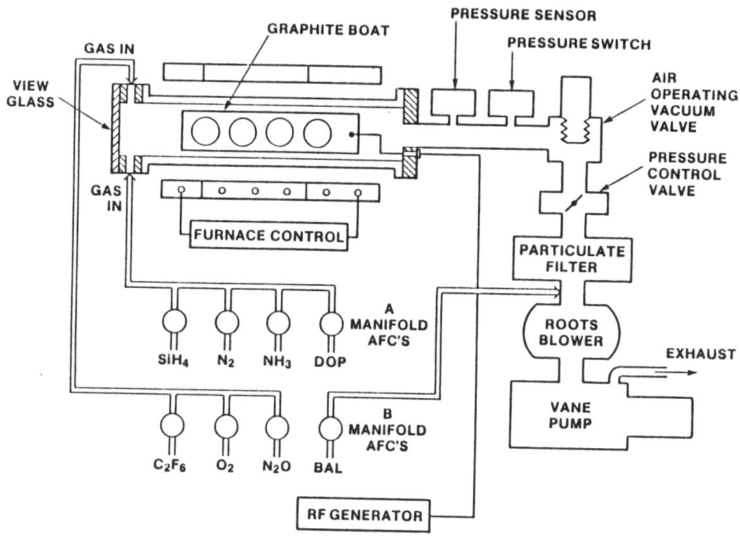

Figure 21: Schematic of hot wall plasma deposition system, PWS-450. Pacific Western Systems, Inc.

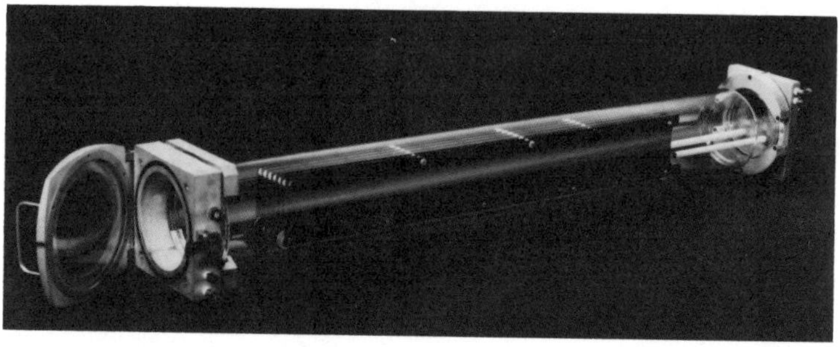

Figure 22: Plasma-enhanced CVD reaction chamber, PWS-450. Pacific Western Systems, Inc.

The major advantage of the hot tube approach is in wafer throughput. A typical wafer load would be seventy 3" wafers, as contrasted to the 42-wafer load in the AMP-3300. The advantage is not quite 2:1, as the hot tube wafer boat cools off when it is withdrawn to replace finished wafers. Time to reheat this structure lengthens processing time.

As in the LPCVD reactor discussed earlier, allowing the electrode structure to touch the tube wall as it is inserted leads to considerable particle contamination. Therefore, cantilever loaders are available here also, and a typical unit on an ASM reactor is shown in Figure 23. Again, in contrast to the AMP-3300, this system is operated under computer control. Automated handling of wafers is more difficult to achieve, and is not generally available.

Figure 23: Cantilever loading system for plasma-enhanced CVD system. ASM, Inc.

6.6 NEW CONCEPTS

All of the reactor systems described so far have shortcomings of one sort or another. They may have excellent throughput, but introduce too many particles into the process. They may be awkard to disassemble and clean, leading to substantial downtime. Uniformity on individual wafers has historically been adequate at ± 5%. Today, uniformities of ± 1 to 2% are being required because of more stringent process requirements, and this is hard to achieve over all the wafers in a large batch load. Uniformity of dopant on a wafer is also a concern.

As a result of the ever tighter specifications being imposed on CVD systems, a number of new generation systems have been introduced recently. We will review several of these, although none has yet achieved any substantial commercial acceptance.

6.6.1 Hot Wall Cross-Flow Reactor

One of the difficulties with the traditional LPCVD hot tube reactor is depletion of reactant gases as they flow down the length of the tube. To overcome this problem, designers either ramp up the temperature at the tube back end to increase deposition rates, or introduce the gases in a distributed fashion along the tube length. Neither solution is ideal, so several years ago Anicon introduced a new hot wall configuration. In this system, a hot quartz bell jar is used to provide the uniform temperature ambient for deposition.

The reaction chamber is also quartz, and is contained within the larger bell jar. Since the reaction chamber is operated at low pressures, a ballast gas occupies the space between the bell jar and the reaction chamber.

Reactant gases enter the reaction chamber along its vertical axis, and are introduced at the top. The wafers are placed vertically in cassettes with typical close spacing. The gas then flows parallel to the wafer surfaces and leaves the system. Therefore, the only reacting gases a wafer sees are fresh ones, and depletion is not an issue.

Another feature of this approach is a claimed lower particulate count that the wafers are exposed to. If the longer flow path in an LPCVD furnace promotes gas phase nucleation and stirs up particles, then this system should minimize such effects.

In the traditional LPCVD hot tube reactor, cleaning the tube required removing it from the furnace, which can be quite a job. In the Anicon system, the claim is that the quartz reaction chamber can be replaced in 15 minutes.

Anicon claims this reactor can process up to one hundred 5" wafers at one time. Temperatures up to 740°C are available, and pressures of 250 mTorr to 10 Torr can be run. All LPCVD processes that were described earlier in Chapter 3 can be executed on this system.

6.6.2 Cold-Wall Thermal Systems

All of the remaining new CVD reactor systems are cold wall reactors carrying out thermal or plasma-enhanced CVD processes. They are being developed to deposit a variety of films, but each system is initially targeting a particular material.

The first of these was developed several years ago to deposit tungsten and tungsten silicide films. A diagram of the reactor chamber is shown in Figure 24. It is based on the axisymmetric cold-wall hexode reactor heated from within, described in Chapter 1. Since there are eight wafers inside the reactor at one time, it is a batch reactor, even though there are eight gas injection ports. If the susceptor did not rotate during deposition, one could characterize it as a single chamber with eight single-wafer stations.

The obvious question to ask is why not run this process in a hot tube and treat 50 to 100 wafers at a time rather than eight? The explanation is that the WSi_2 process (WF_6 + SiH_4 → WSi_2 + ...) proceeds very rapidly so that when reasonable deposition rates are achieved, the process is diffusion controlled. As we recall, the hot tube batch reactor only works when the process can be operated in a surface-controlled regime.

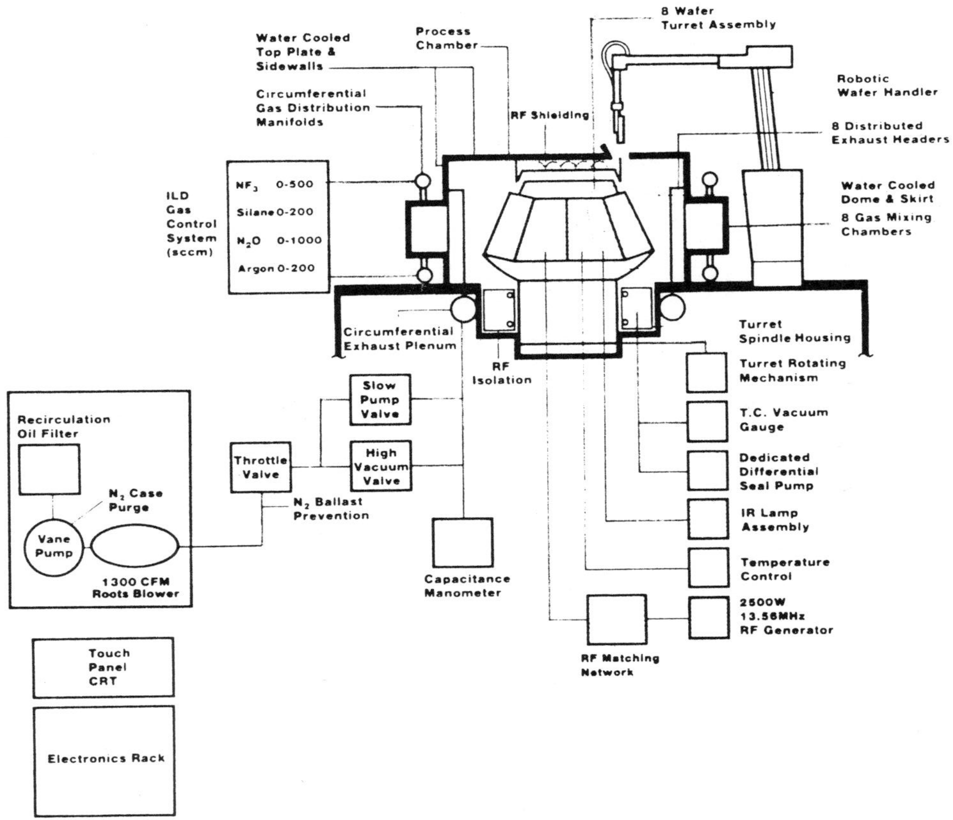

Figure 24: Genus tungsten CVD reactor system.

A second approach to a cold wall system is the single-wafer CVD reactor developed by Varian-Torrex. A schematic of the reaction chamber is shown in Figure 25. Again, tungsten silicide is deposited in this cold-wall reactor. Other conducting films such as blanket and selective tungsten can also be deposited.

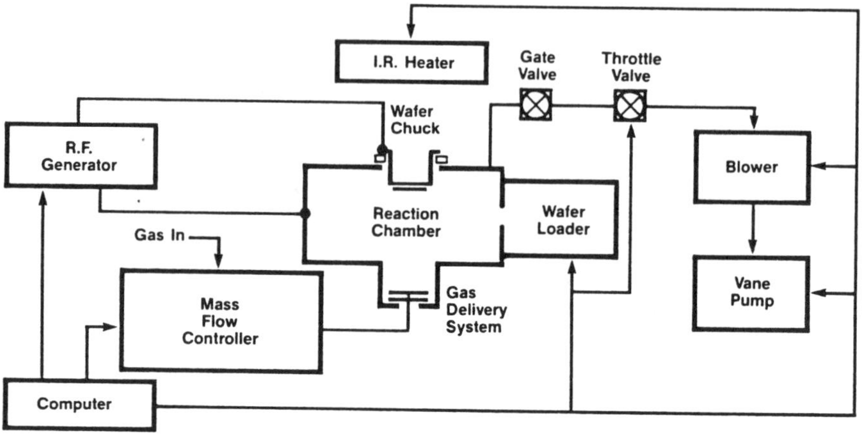

Figure 25: Varian-Torrex 5101 single-wafer CVD reactor system.

In comparison to the Genus reactor, this system holds the wafer upside down to minimize any particulate on the wafer. Also, since this is a single-wafer machine, a loadlock is provided to ensure that the reaction chamber is never opened to the atmosphere. Attempts to provide this feature on a batch reactor are difficult and expensive, due to the size of the chamber needed. Heating is done in a way similar to the Genus system. High-intensity lamps shine on the back of a chuck to heat it to processing temperature.

A final point should be made concerning the single-wafer CVD reactor concept. This approach only makes sense if each wafer can be processed in 1 to 2 minutes, so reasonable throughput can be achieved. In many applications, conducting films can be thin, ~2000 Å, so deposition rates of 1000 to 1500 Å/min would be suitable. Such rates are not unreasonable, for example, for WSi_2 films.

The remaining system is a plasma-enhanced CVD system for the low-temperature deposition of low hydrogen content silicon nitride. The system is shown in Figure 26, and a schematic of the reaction chamber in Figure 27. As can be seen, this reactor is a batch system where the wafers are placed in a square array.

In this reactor, N_2 is introduced into a number of small glow discharge chambers. At the same time, silane flows into the chamber adjacent to but not in the discharge chambers. In the latter, the N_2 dissociates, and because of its long recombination time, N atoms are available to react with the silane on the wafer surface. Because of this pre-ionization and dissociation of N_2, it is not necessary to heat the wafer to promote the reaction at reasonable rates. In an arrangement such as this, there will be little ion bombardment of the wafer during deposition. If such bombardment were desired (i.e., enhance compressive stresses), a second electrode can be powered, as shown, to create a plasma around the substrates.

As noted earlier, this is the only system on the market that can deposit good quality silicon nitride films at room temperature. As low-temperature processing becomes more valuable, this approach will attract more and more attention.

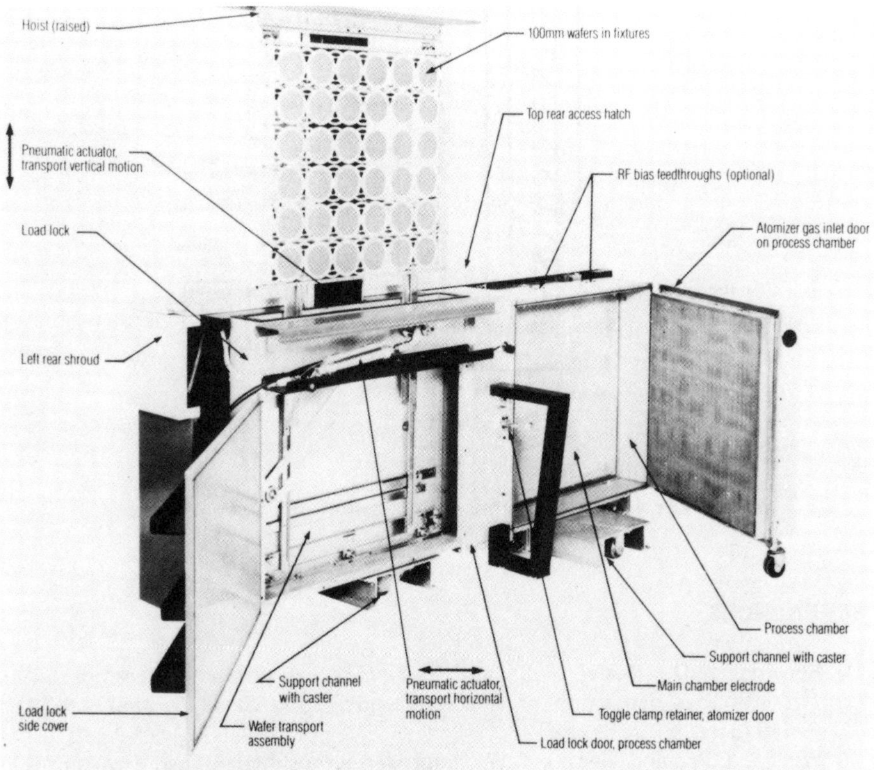

Figure 26: Low-temperature PECVD reactor system—Ionic Systems, Inc.

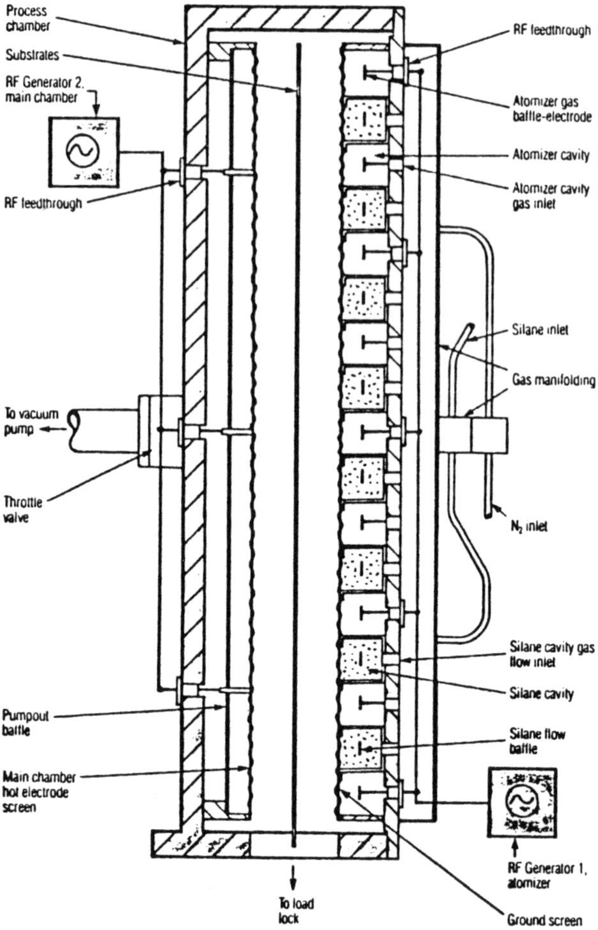

Figure 27: Reaction chamber schematic—Ionic Systems, Inc.

REFERENCES

1. Benzing, W.D., Rosler, R.S., and East, R.W., A production reactor for continuous deposition of silicon dioxide. *Solid State Technol.* 16:37 (1973).
2. Winkle, L.W., and Nelson, C.W., Improved atmospheric-pressure chemical-vapor-deposition system for depositing silica and phosphosilicate glass thin films. *Solid State Technol.* 24(10):123 (1981).
3. Sherman, A., Design of plasma processing equipment. To be published.

7

Film Evaluation Techniques

7.1 INTRODUCTION

Throughout all the preceding chapters, we have discussed thin films that can be created by chemical vapor deposition in terms of their physical and chemical attributes. However, we did not explain how we secured the necessary physical or chemical information. For example, we discussed film deposition rates many times, but did not explain how we knew the film thickness after a specified amount of time. Similarly, when we spoke of the stoichiometry of deposited composite films, we did not indicate how we determined their chemical composition.

In the present chapter, we will attempt to correct this oversight. The first half of the chapter will review the many techniques whereby we measure the physical nature of the film we have deposited. The second half will cover the chemical composition of the film, both in bulk (average over film thickness) as well as how it varies through the film thickness.

Since the measurement techniques for thin films from several microns down to several hundred Angstroms thick are quite sophisticated, it was felt that their detailed description would be better left to a separate chapter. In this way, they can be dealt with in some detail without interfering with the study of the various CVD techniques.

7.2 PHYSICAL MEASUREMENTS

In this section, we will discuss those techniques one uses to evaluate the physical characteristics of the thin films we can deposit. We specifically defer questions as to the chemical nature of the film.

7.2.1 Thickness

The measurement of film thickness can be a fairly simple measurement or it can be quite complex, depending on the nature of the film. The most direct technique is the measurement of the step height when a portion of the deposited film is etched away. This is done by electronically tracking the position of a mechanical stylus as it is traversed across the step. Such a surface profilometer is illustrated in Figure 1.

A typical surface profile is shown on the video display. Vertical resolution of 5 Å and horizontal resolution of 400 Å is claimed. As long as the deposited film can be etched off the substrate without etching the substrate, this technique can be used for any thin film. Its primary utility is for R&D studies, as it is clearly not a production technique. The only film for which it is not suited is an epi silicon film on a single-crystal silicon substrate. A technique for measuring the thickness of these films will be described in the section on Infrared Spectroscopy.

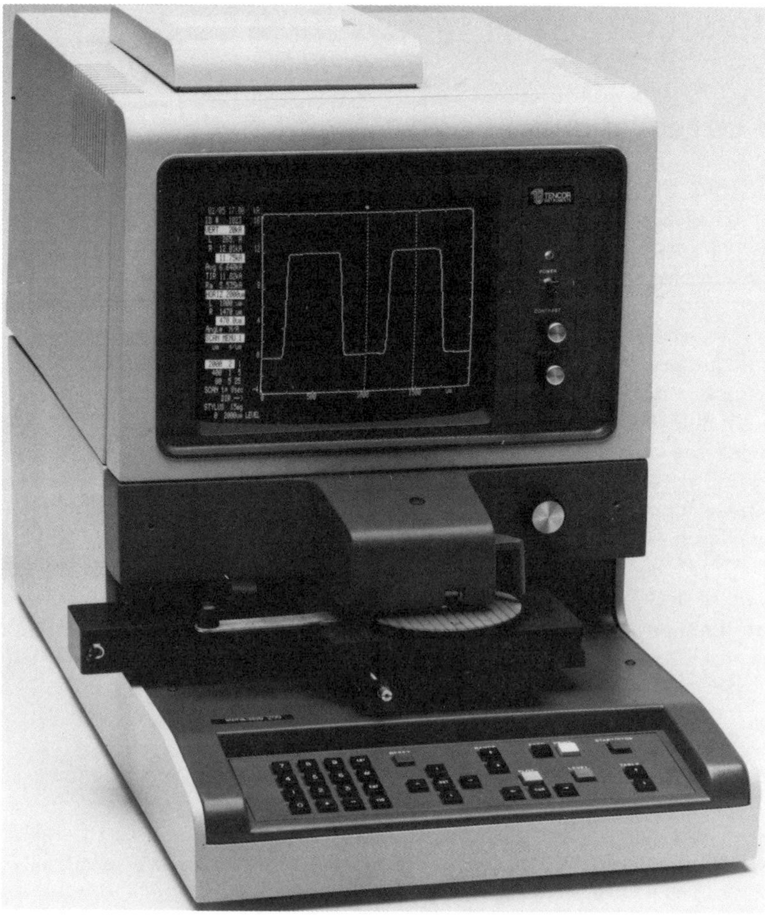

Figure 1: Computerized surface profilometer, Alpha-Step 200 Tencor Instruments.

As long as the film is not reflective (i.e., specular aluminum) and is deposited on a reflective substrate (i.e., SiO_2 on silicon), optical techniques are available. It was recognized early that the color of a thin film could be correlated to its thickness. Although not very precise, such information is very useful for quick evaluation in the laboratory. For example, silicon dioxide films on silicon substrates can be evaluated with the data of Table 1. In fact, one of the more useful aspects of this technique is that one can make rapid judgements as to film uniformity.

Going beyond this simple qualitative technique, the thickness of films can be measured by a polarizing spectrometer or "ellipsometer." This is an instrument whose operation is based on the fact that elliptically polarized light changes its polarization upon reflection from a thin transparent film on a reflecting substrate. The ellipsometer creates an elliptically polarized monochromatic light beam, and then evaluates the light beam on reflection from a thin film. The essential ingredients from an ellipsometer are shown in Figure 2.[2] A monochromatic beam of light (today most often from a laser) passes into a polarizer where it becomes plane polarized. It then passes through a compensator which converts it into an elliptically polarized light beam. After reflection from the substrate/thin film, it passes through an analyzer. If it had been converted back to plane polarized when it had been reflected, then it would be possible to rotate the analyzer to find a true minimum intensity. The technique then is to adjust the polarizer until the reflected light is plane polarized. The analyzer is rotated to determine the position corresponding to a minimum in light intensity. This information, along with a theoretical model of the optical process almost 100 years old, permits a calculation of the film thickness.

With the advent of modern computing capabilities, ellipsometers have been automated and have proven useful in production settings. Originally, this technique was found most useful for the evaluation of dielectric films deposited on silicon substrates. Today, more sophisticated instruments such as the one shown in Figure 3 can be used to measure a wide variety of thin films on many different substrates. Even metal films can be measured if they are less than 500 Å thick.

Finally, we should note that in addition to film thickness, the index of refraction of the film can be determined and used to obtain chemical information about the film. This aspect will be discussed in Section 7.3.1.

Another instrument widely used to measure film thickness is a spectrophotometer that operates over the visible light (4800 to 8000 Å) wavelength range. This instrument essentially quantifies the qualitative evaluation of film color mentioned earlier. A commercial instrument operating on this principle is shown in Figure 4.

Light reflected from the thin film is passed through the optical microscope onto a dispersive grating. The grating is then mechanically rotated so that the light spectrum is passed over a thin slit. The intensity of light passing through the slit is measured by a photointensity meter and recorded by the computer. In this way, the most intense frequency (color) is determined. This information, plus knowledge of the index of refraction, allows the film thickness to be determined.

Table 1: SiO_2 Thickness vs. Color[1]

FILM THICKNESS (MICROMETERS)	*ORDER (5450 A)*	*COLOR AND COMMENTS*
0.050		tan
0.075		brown
0.100		dark violet to red violet
0.125		royal blue
0.150		light blue to metallic blue
0.175	I	metallic to very light yellow green
0.200		light gold or yellow—slightly metallic
0.225		gold with slight yellow orange
0.250		orange to melon
0.275		red violet
0.300		blue to violet blue
0.310		blue
0.325		blue to blue green
0.345		light green
0.350		green to yellow green
0.365	II	yellow green
0.375		green yellow
0.390		yellow
0.412		light orange
0.426		carnation pink
0.443		violet red
0.465		red violet
0.476		violet
0.480		blue violet
0.493		blue
0.502		blue green
0.520		green (broad)
0.540		yellow green
0.560	III	green yellow
0.574		yellow to "yellowish"*
0.585		light orange or yellow to pink borderline
0.60		carnation pink
0.63		violet red
0.68		"bluish"**
0.72	IV	blue green to green (quite broad)
0.77		"yellowish"
0.80		orange (rather broad for orange)
0.82		salmon
0.85		dull, light red violet
0.86		violet
0.87		blue violet
0.89		blue
0.92	V	blue green
0.95		dull yellow green
0.97		yellow to "yellowish"
0.99		orange
1.00		carnation pink
1.02		violet red
1.05		red violet
1.06		violet
1.07		blue violet
1.10		green
1.11		yellow green
1.12	VI	green
1.18		violet
1.19		red violet
1.21		violet red
1.24		carnation pink to salmon
1.25		orange
1.28		"yellowish"
1.32	VII	sky blue to green blue
1.40		orange
1.45		violet
1.46		blue violet
1.50	VIII	blue
1.54		dull yellow green

*Not yellow, but is in the position where yellow is to be expected; at times it appears to be light creamy grey or metallic.

**Not blue but borderline between violet and blue green; it appears more like a mixture between violet red and blue green and overall looks greyish.

NOTE: Above chart may also be used for Vapox, Silox, and other deposited oxide films. For silicon nitride films, multiply film thickness by 0.75.

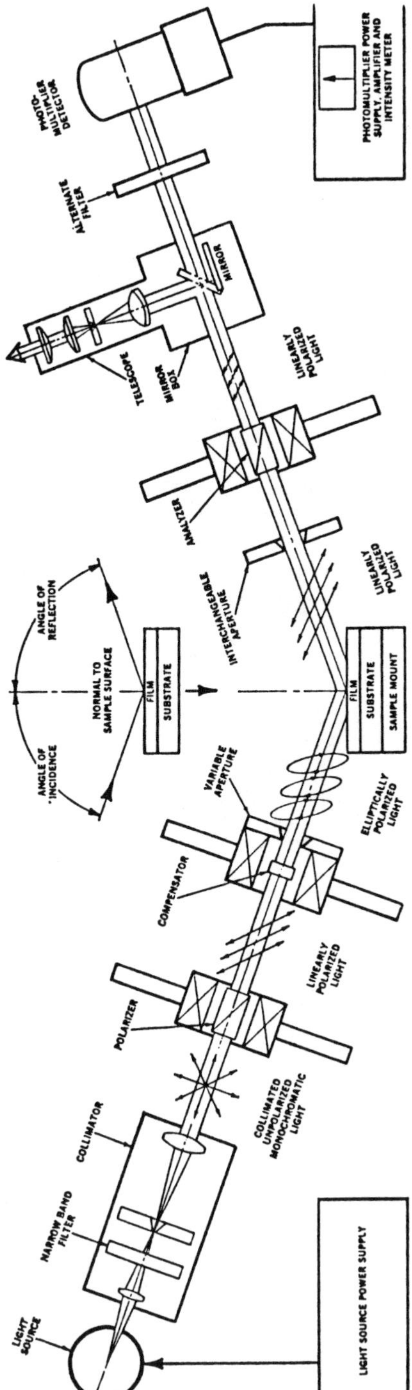

Figure 2: Elements of typical ellipsometer.[2]

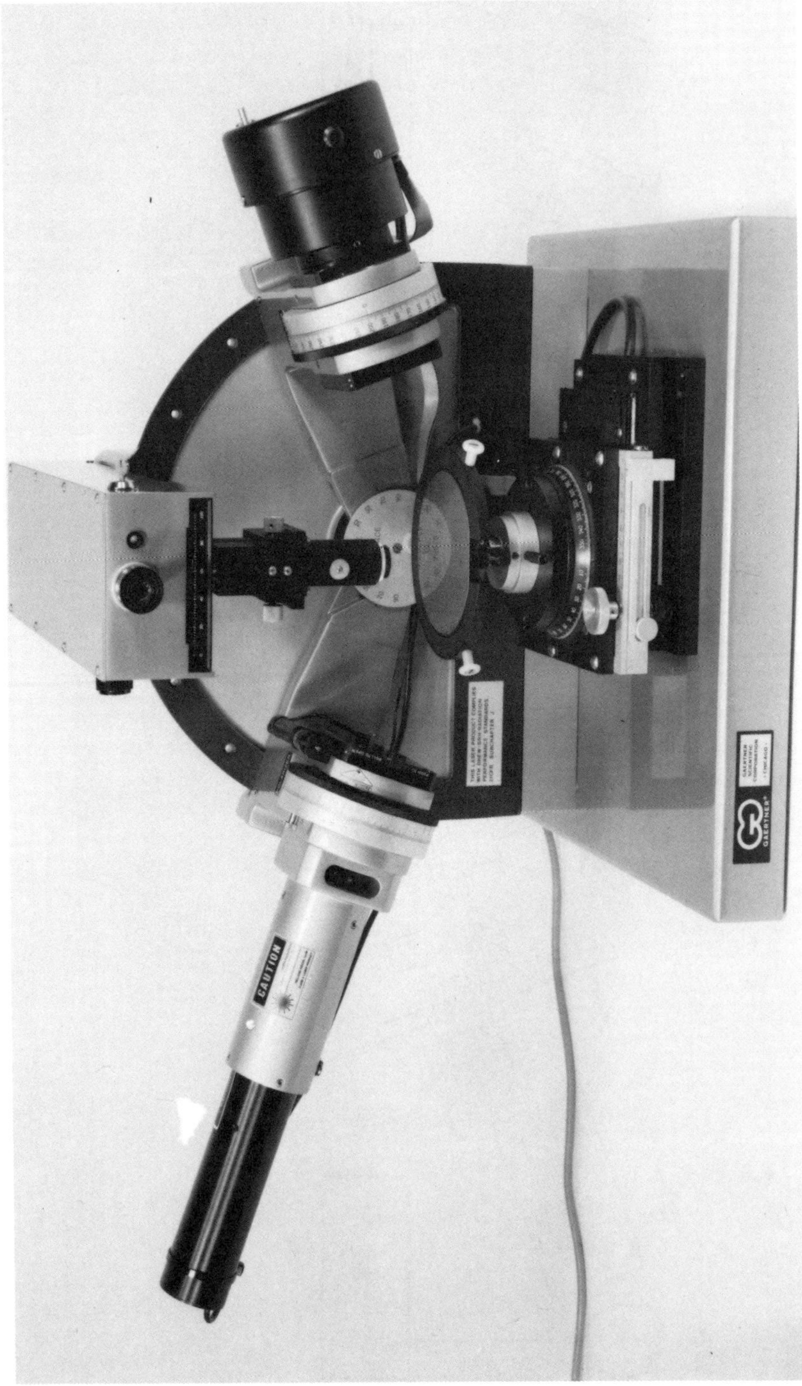

Figure 3: Commercial automated ellipsometer—Gaertner Scientific Corporation.

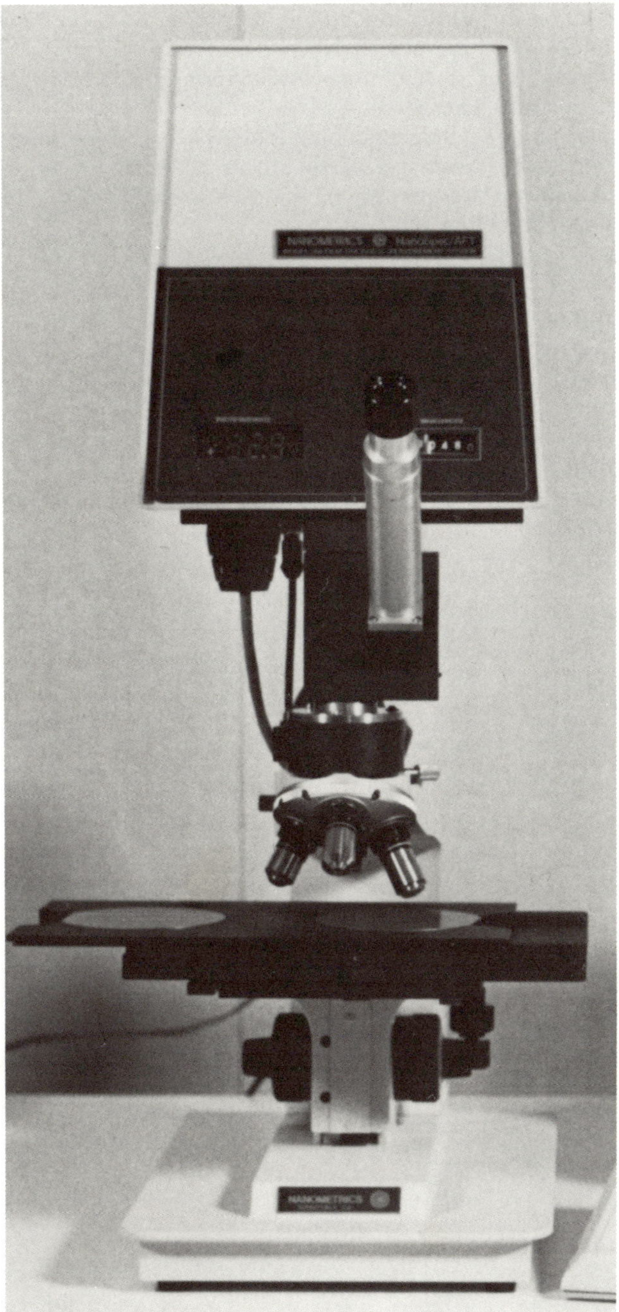

Figure 4: Automated spectrophotometer–NanoSpec/AFT 200 Nanometrics, Inc.

The only disadvantage of this instrument is the requirement that the index of refraction be known. Particularly for plasma-enhanced films, the index of refraction can vary considerably, depending on deposition conditions. However, for a production process where the index of refraction is well known, and the primary issue is film thickness uniformity, the Nano Spec can provide information very quickly.

The thickness of conducting films on semi-conducting substrates can be inferred from sheet resistance measurements provided the resistivity of the thin film is known. Measurement of the sheet resistance of conducting films will be reviewed in Section 7.2.3.

7.2.2 Stress

When a thin film of one material is deposited on a stress-free substrate of another material, one may find the thin film to be under tensile or compressive stress. The nature of this stress is typically evaluated by depositing the thin film on a silicon wafer and measuring its deflection. As illustrated in Figure 5, a tensile stress will deflect the substrate downward, and a compressive stress will deflect it upward.

A formula for the stress has been developed assuming the substrate to be a circular plate.[3] It is

$$\sigma = \left(\frac{\delta}{r^2}\right)\left(\frac{E}{3(1-\nu)}\right)\left(\frac{t_s^2}{t_f}\right) \tag{1}$$

where

σ = stress (dynes/cm^2)

δ = disc deflection (cm)

ν = substrate Poisson's ratio

E = substrate Young's modulus

t_f = film thickness (cm)

t_s = substrate thickness (cm)

r = disc radius (cm)

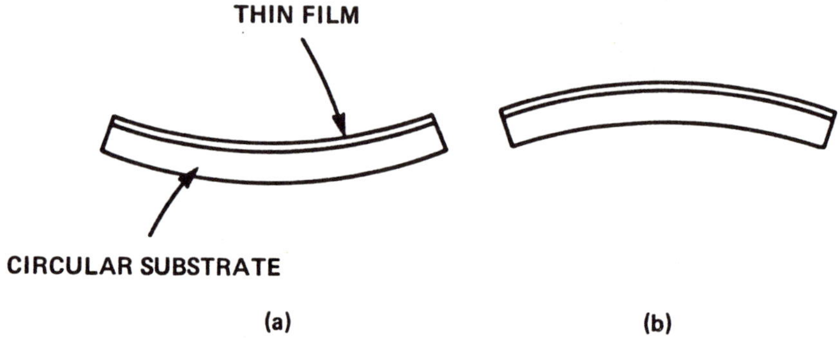

Figure 5: Thin film stress: (a) tensile, (b) compressive.

A variety of techniques may be used to measure the wafer center deflection. The simplest is to place the wafer on the stage of an optical microscope and calibrate the vertical adjustment. Then compare the in-focus vertical position at the wafer edge to that at the wafer center.

One commercial wafer deflection gauge is available, and is sketched in Figure 6. The degree of light reflection is used to indicate the amount of wafer deflection. The only difficulty with this technique occurs when relatively low stress films are measured. For normal films (i.e., thermal CVD silicon dioxide) and a stress of 10^9 dynes/cm^2, a typical 100-mm silicon wafer (0.62-mm thick) with a 1-μm thick film will deflect ~10 μm at its center. The Ionic Systems gauge claims a 0.03-μm sensitivity, so the typical stress can be measured readily. For smaller stresses ~10^8 dynes/cm^2, it may be useful to use a thinned wafer to make deflection measurements.

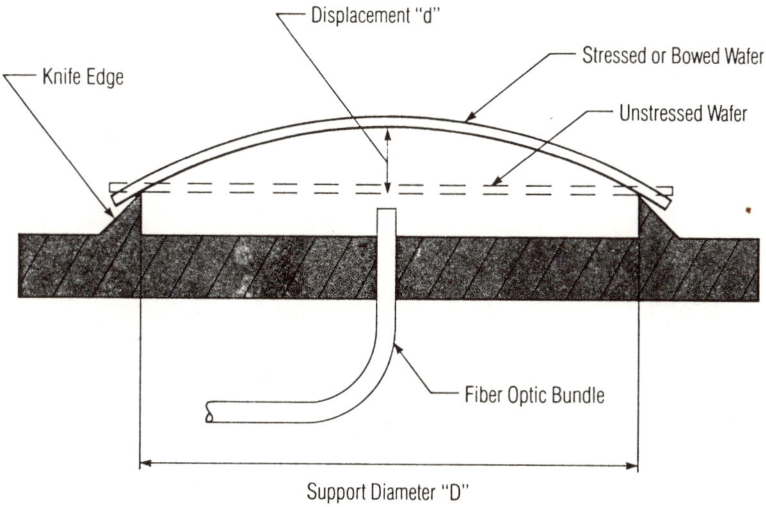

Figure 6: Wafer deflection gauge—Ionic Systems, Inc.

There are a number of subtle effects that have to be considered when making thin film stress measurements on silicon wafers.[4] First of all, the crystal orientation of the wafer influences the resulting stress. The same thermal CVD silicon dioxide film thickness on the same substrate indicates larger tensile stresses on (100)-oriented wafers as compared with (111)-oriented wafers.

It has also been shown that the stress in a deposited film will change with time, depending on how the wafer is stored. Figure 7 shows the deflection (stress) as a function of time for a wafer stored in a dry box versus a wafer stored at 100% humidity.[4] When the second wafer was returned to a dry ambient, its deflection returned to its original value. Clearly, stress will depend on the ambient conditions under which wafers are measured.

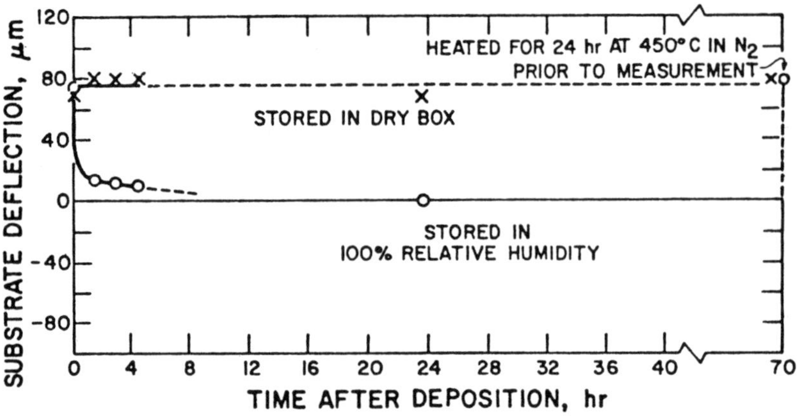

Figure 7: Wafer deflection vs. storage ambient and time.[4]

There is also evidence that film stress depends on film thickness, although there are conflicting reports. Therefore, it is prudent to measure stress using films equal in thickness to those eventually to be used. For very thin films and small stresses, quite small deflections may have to be measured.

7.2.3 Sheet Resistance

The electrical resistivity of thin films of conducting material deposited on insulating substrates is an important quantity to measure. Consider a rectangular conductor, as shown in Figure 8. Current flows through this conductor and a potential difference V exists between its ends. The resistance of this macroscopic segment is, according to Ohm's Law:

$$R = \frac{V}{I} \quad (2)$$

If we are interested in the local resistivity of a material, we express Ohm's Law in terms of the local electric field and the local current density. Then:

$$\rho = \frac{E}{i}, \quad (3)$$

where E is in volts/cm, i is in amps/cm^2, and ρ is in Ω-cm.

We then write for resistivity,

$$\rho = \frac{V/\ell}{I/(bd)} = R \cdot \frac{bd}{\ell}. \quad (4)$$

If we choose $\ell = b$, then we can define a sheet resistance which is

$$R_S = \frac{\rho}{d} \text{ (ohms per square, or } \Omega/\square) , \quad (5)$$

which is independent of the size of the square as long as a square is used to make the measurement. In this way, a measurement of sheet resistance leads directly to the local film resistivity.

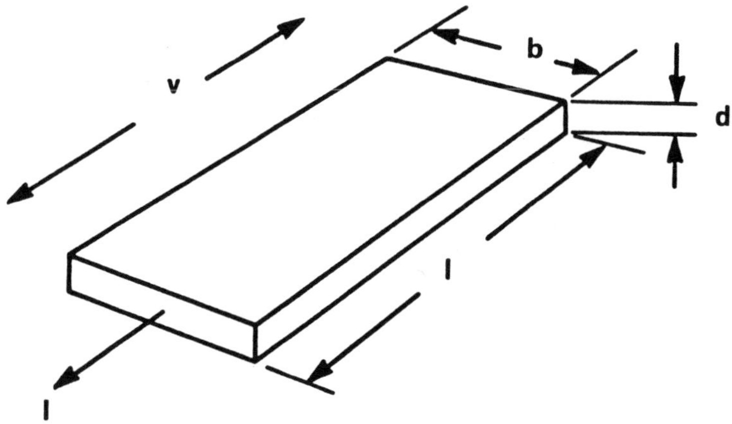

Figure 8: Thin film conductor.

Sheet resistance can be measured with a four-point probe. The probes may be in line or in a square pattern, as shown in Figure 9. In either configuration, a constant current I is passed through two of the probes, and the voltage difference between the other two is read. Provided the conducting layer is thin ($t \leqslant 0.60\ d$), the sheet resistance can be calculated from

$$R_S = 4.53\frac{V}{I} \quad \text{In-line Probes}^{5}$$

(6)

$$R_S = 9.06\frac{V}{I} \quad \text{Square Probes.}^{6}$$

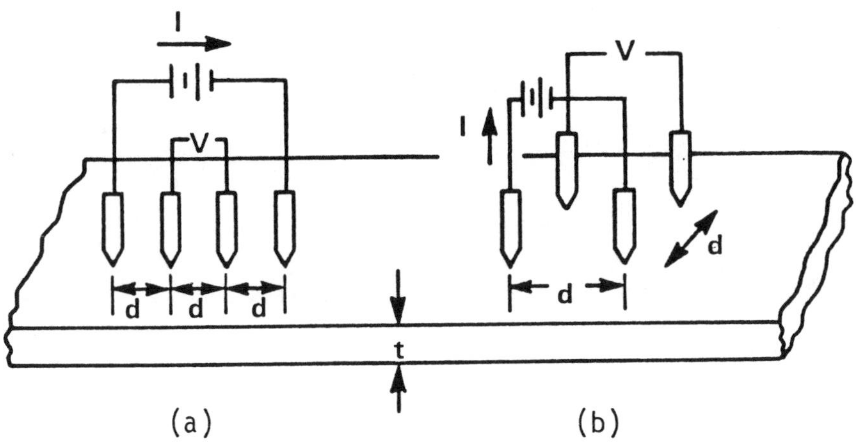

Figure 9: Four-point probes to measure sheet resistance. (a) in-line; (b) square.

Since most conducting films of practical interest will be less than 10,000 Å thick, reasonable probe spacings (d > 1.5 μm) allow the use of the above relations. Equation (5) then enables a determination of the average film resistivity.

A commercial instrument that automatically measures sheet resistance is shown in Figure 10, a typical sheet resistance map of a wafer is shown in Figure 11.

This instrument uses a four-point in-line probe with spacing between probes of 0.040". When the probe is near the wafer center, the expression given in Equation (6) is valid. Near the wafer edge, however, this relation can only be used with a geometric correction factor. Instead of this, a "configuration switching" technique is used to automatically compensate for any such geometric factors. In this case, the voltage is first read on the inner two probes when current is passed through the outer two. Then, the current is passed through the first and third probes, and a voltage read between the second and fourth probes.

Figure 10: Instrument to measure sheet resistance, Prometrix Corp. Omnimap® RS50.

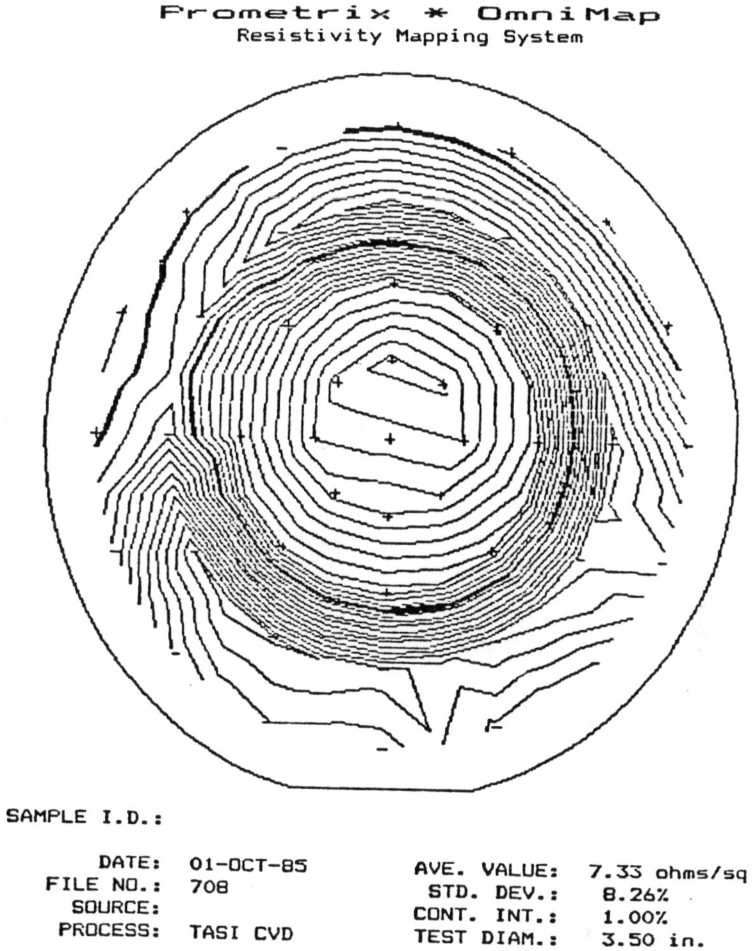

Figure 11: Sheet resistance map of silicon wafer with 3000 Å film of tantalum silicide.

A geometrical factor, k, can then simply be calculated from a ratio of these two voltages and the sheet resistance calculated from

$$R_S = 4.53(k)\frac{V}{I}\ , \tag{7}$$

where V and I are obtained when current is passed through the outer probes.

In general, many points are read on a wafer surface, and a contour map of sheet resistance is produced, as was illustrated in Figure 11. Here the heavy line represents the average value, and contour lines (+ or –) indicate a percent change from the average.

For conducting films of known resistivity, the above information can be used to generate a map of film thickness.

7.2.4 Visible Defects

When depositing amorphous or polycrystalline films, defects take the form of particles incorporated into the film either from dust particles or gas phase nucleation. For epitaxial films, as discussed in Chapter 3, there can be a variety of crystallographic defects (stacking faults, slip lines, spikes, etc.). In the as-deposited films, these defects are not always readily visible. Therefore, epi films are frequently etched with a "preferential" chemical etchant to more clearly delineate whatever faults there may be. A frequently used etch is the "Wright" etch.[7] It is a mixture of HF, HNO_3, CrO_3, $Cu(NO_3)_2$, acetic acid and deionized water. Stacking faults occurring in epi layers with a 5-minute Wright etch are shown clearly in Figure 12.

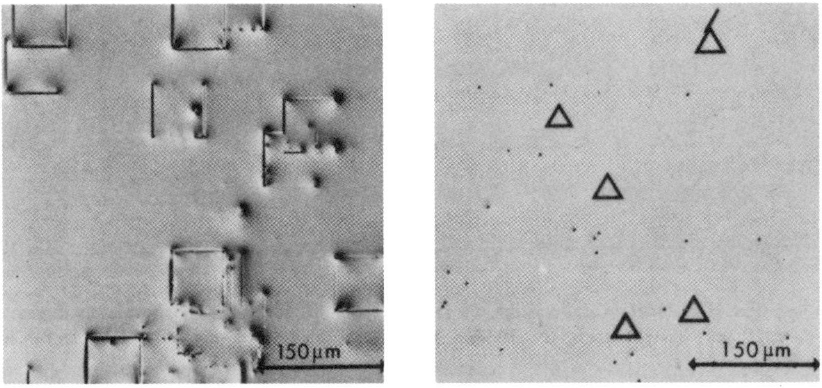

Figure 12: Stacking faults in epi layers: (a) (100) orientation; (b) (111) orientation.[7] Reprinted by permission of the publisher, The Electrochemical Society, Inc.

Obviously, defects can be seen with an optical microscope. What is needed is a quick way to count different types of defects on each wafer processed. For this purpose, a commercial computerized unit has been developed that can distinguish between point, line and area defects. The measurement and wafer-handling unit is shown in Figure 13.

The system uses a helium-neon laser to scan a wafer surface. Light scattered by defects is collected and amplified, and the resulting photomultiplier signals reveal the location and nature of the defect. Particles as small as 0.3 micron can be detected.

Figure 13: Surface defect detector–Surfscan 160, Tencor Instruments.

7.2.5 Morphology–SEM/TEM

Throughout the earlier chapters, we have seen thin film surface morphology illustrated by scanning (SEM) and transmission (TEM) electron microscope pictures. The SEM is the only technique available to examine thin film surfaces with submicron size features, because optical microscopes are limited by the wavelength of light. Even for larger features, the SEM is valuable because its depth of field is much greater (500X) than that of an optical microscope. Resolutions of 25 to 100 Å are possible with a well-maintained and carefully operated system.

The SEM operates by scanning a focused electron beam over a surface and sensing the secondary electrons emitted from the surface. A TEM on the other hand operates like an optical microscope by passing an electron beam through a sample. The electron beam is then magnified by electromagnetic lenses up to 1,000,000 times to form an image. Sample preparation is the most time consuming part of using TEM for studying the structure of CVD thin films. The procedure typically followed is to thin a sample by ion milling the back of the area of interest, until it is thin enough to pass the TEM electron beam. Film thickness on the order of 1000 Å is necessary.

7.3 CHEMICAL MEASUREMENTS

In the following sections, we will deal with only some of the more common techniques used to evaluate the chemical nature of CVD films. We will be interested in the composition of the thin films, both as an average over the film thickness as well as a function of position in the film. We will also consider the chemical state of the atoms in terms of the bonds they can form within the film.

7.3.1 Refractive Index—Ellipsometry

The ellipsometric technique described earlier has the unique feature that the index of refraction can be determined independently of the film thickness. Then, knowledge of this index can be used to infer the chemical composition of a film. For example, thin silicon dioxide films have an index of 1.46, while silicon nitride films have a value of 2.0 typically. Now, when either of these films are deposited by PECVD techniques, their stoichiometry can vary depending on deposition conditions. It turns out that this variation in stoichiometry can be related to the measured refractive index. Accordingly, measurements of the refractive index can be used as an approximate guide to film stoichiometry.

7.3.2 X-Ray Spectroscopy

Within this technique, we include EDX (energy dispersive x-ray analysis), WDX (wavelength dispersive x-ray analysis), and XRF (x-ray fluorescence analysis). In all of these, x-rays emitted from a sample are analyzed. In one case, they are created by bombarding the sample with x-rays (XRF), and in the others, they are created by high energy electron beam as in an SEM (EDX, WDX).

The method involves the absorption of a high-energy x-ray or the energy of an energetic electron by an atom in the sample. This atom then de-excites by emitting x-rays. These emitted x-rays have energies or wavelengths typical of the bombarding x-rays or electrons and of the binding energy of the excited bound electron which relaxes to its original ground state. The x-rays emitted by the sample can be detected individually, and catalogued (by computer) according to energy. In this way, all elements heavier than fluorine can be detected simultaneously. Concentrations of 1% are detectable by this approach. If the wave lengths are analyzed, then sensitivity can be increased to the ppm range, and elements down to boron, carbon and oxygen can be detected.

Use of these x-ray techniques is only applicable to films as thick as 1 to 3 microns, since the emitted x-rays come from this depth in a sample (as illustrated in Figure 14). For a 2000 Å layer of tungsten on silicon, for example, we would clearly detect tungsten and silicon. However, there would be no way to know if any oxygen detected came from the substrate or the thin tungsten film.

Finally, when an EDX unit is attached to a SEM, it becomes possible to evaluate surface film compositions with high resolution, and produce concentration maps corresponding to the area viewed. Therefore, this is a frequently found addition to a typical SEM installation.

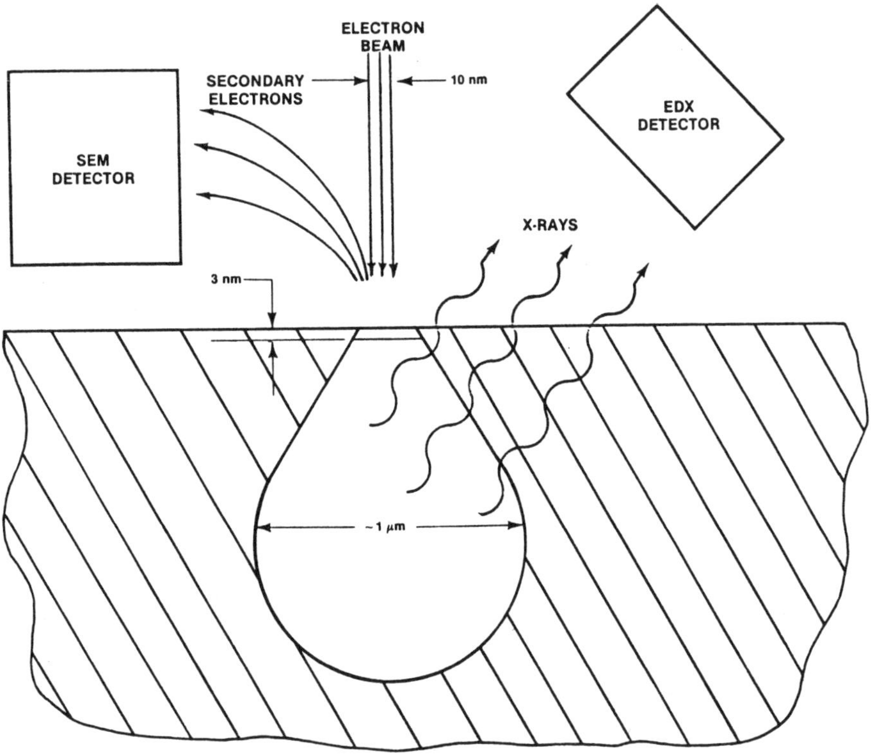

Figure 14: Simultaneous use of SEM and EDX.[8]

7.3.3 Dopant Distribution

As noted earlier in Chapter 3, epitaxial silicon films deposited by CVD can be affected by autodoping. If diffusion of the doping species is excessive, the film is not a useful one. Therefore, quite a lot of effort has been spent to accurately measure the distribution of dopant through the film thickness.

One technique is referred to as the "spreading resistance" method. In this procedure, a wafer is fractured and the edge containing the film is beveled, as shown in Figure 15. Then, a two-point probe is used to measure resistivity at a sequence of points traversing the interface between the substrate and the epi film. By relating the local resistivity to carrier concentration, one is able to deduce the concentration of dopant atoms over the epi layer. This technique is effective for even highly-doped layers.

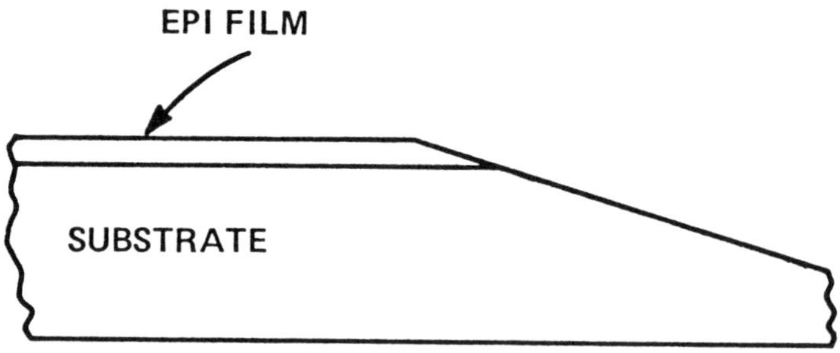

Figure 15: Beveled segment of epi wafer.

For less highly-doped epi films, one can use the C–V method. In this case, use is made of the fact that a Schottky semiconductor diode has a voltage-dependent capacitance. In other words, when such a diode is reverse biased, a depletion layer forms which then has a capacitance determined by the depth of this layer (w) as well as the doping (N) at its edge. The doping profile can be determined from the following relations.[9]

$$N(w) = -\frac{C^3(v)}{e\epsilon A^2 \frac{dC(v)}{dv}} \tag{8}$$

and

$$w = \frac{\epsilon A}{C(v)}, \tag{9}$$

where ϵ is the permittivity of silicon, A is the junction area, e is the electronic charge, C is the capacitance, and v is the bias voltage.

For some doped layers, such as a low-dose ion implant for a MOS transistor, this procedure does not reveal the entire doping profile.[10] In this case, a MOS structure is examined rather than a Schottky diode. Typically, a bare silicon wafer is oxidized, and then aluminum dots are sputtered on to form many MOS structures. When a MOS device is examined, Equations (8) and (9) have to be supplemented by

$$\frac{1}{C_m(v)} = \frac{1}{C_{ox}} + \frac{1}{C_s(v)}, \tag{10}$$

where C_m is the measured capacitance, C_{ox} is the oxide capacitance, and C_s is the depletion region capacitance.

As an example of the use of this technique, a silicon wafer lightly doped with phosphorus is doped with additional phosphorus by ion implantation (dose of $3.5 \times 10^{11} cm^{-2}$). A thermal oxide film of 857 Å thickness was initially grown on the wafer. The variation of dopant concentration with depth from the oxide-silicon interface is shown in Figure 16. The rise in dopant close to the interface in the non-implanted sample is due to the usual phosphorus segregation in this region. As can be seen, the implant peak is 1.1 x $10^{16}cm^{-3}$ at a depth of 1700 Å. The doping level returns to that of the original substrate at 7000 Å.

Figure 16: Doping profile for silicon wafer implanted with phosphorus, measured by C–V technique.[10]

7.3.4 Infrared Spectroscopy

Another technique that can be used to determine the chemical nature of a thin film is infrared spectroscopy. Some materials will absorb certain frequencies in the infrared (wavelengths 2 to 25 microns) because of the excitation of vibrational energy transitions in molecular species. In the same way that electronic transitions in atoms can absorb radiation of specific frequencies, the vibration of a molecule (stretching or bending) will have a resonance value, and it will be excited by any radiation of this frequency. Consider the H_2O molecule and its three vibrational modes, as shown in Figure 17. Clearly, each of these vibrational modes has its own resonant frequency, as indicated, and they are all in the infrared range.

Now, when infrared radiation of a particular frequency is passed through a sample containing molecular species, it may or may not be absorbed. If all frequencies are passed through, some will be absorbed to varying degrees, depending on the molecular species involved. For example, a typical spectrum of transmittance (%) versus wave number (cm^{-1}) (wave number = 1/wavelength)

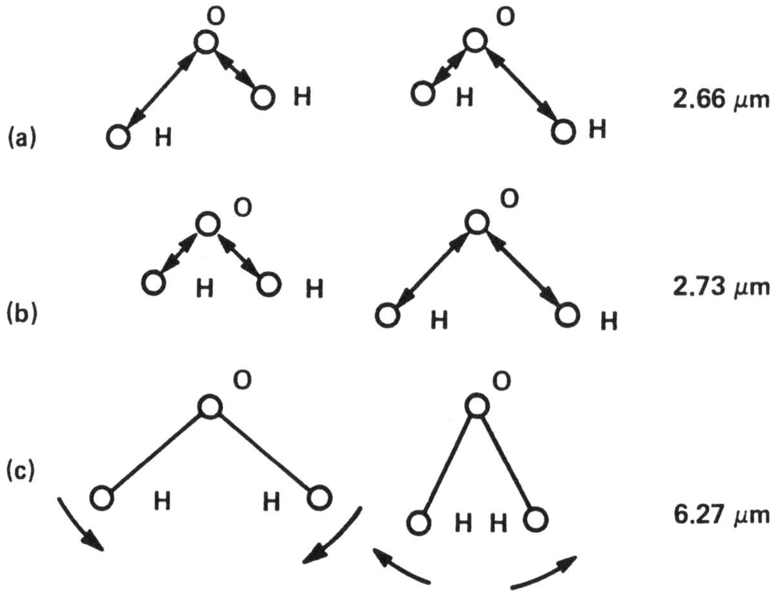

Figure 17: Resonant vibrational modes of the water molecule. (a) asymmetric stretching, (b) symmetric stretching, and (c) scissoring deformation.

for silicon nitride thin film is illustrated in Figure 18. For complex molecules with many vibrational modes, there are many peak absorption frequencies.

There are two types of spectrometers that one can use to generate such spectra.[12] One uses a monochromator to evaluate each frequency in turn. The second uses a Michelson interferometer to examine all frequencies simultaneously, and then a Fourier transform to display the spectrum. The advantage of the latter approach is its greater sensitivity, and the speed with which it can produce a spectrum.

Regardless of how it is obtained, the spectrum can be used to make quantitative estimates of the concentration of molecular species in thin films. Using the Beer-Lambert Law,[12] we can write simply

$$A = E\,C\,L \tag{11}$$

where

A = absorbance = $\log_{10} I_o/I$

I_o = incident radiation

I = Transmitted radiation

E = extinction coefficient

L = path length

C = concentration

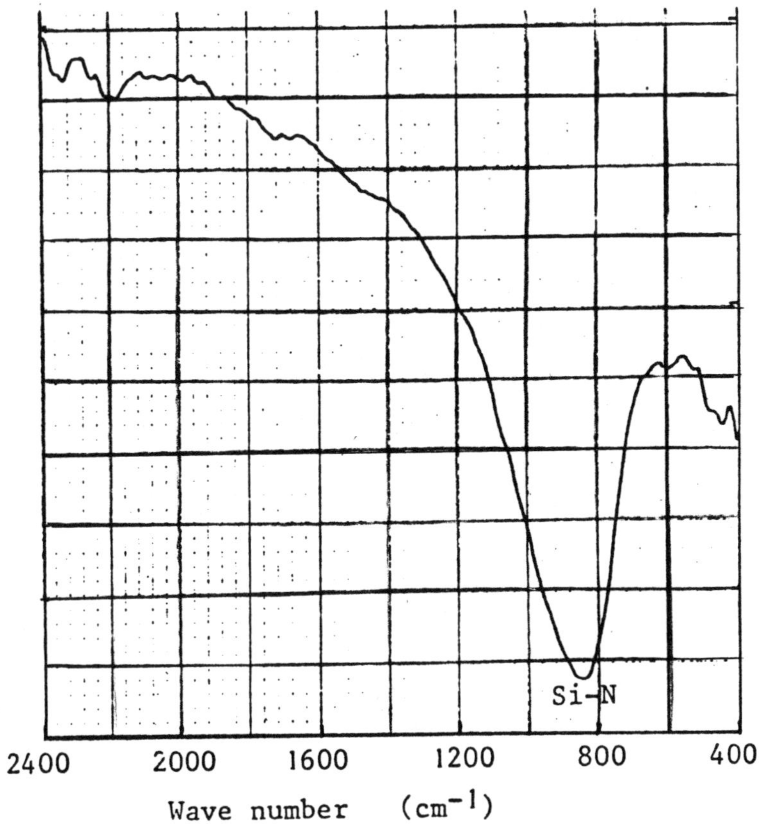

Figure 18: Infrared spectrum for silicon nitride film.[11]

The extinction coefficient is a constant for one substance and one frequency. Then a measurement of A gives a resulting value for C.

Consider the portion of a spectrum near a resonant frequency, as shown in Figure 19. Then, if a tangent is constructed as shown, the absorbance is

$$A = \log_{10} \frac{R}{S}$$

and this is referred to as the "base line density" method. Other methods are based on the area between the tangent and the spectrum trace.

Taking advantage of the speed of a Fourier transform infrared spectrometer (FTIR) and the ability to quantify concentrations, one manufacturer has developed an instrument to be used for quality control of CVD film depositions. A schematic of the instrument is shown in Figure 20.

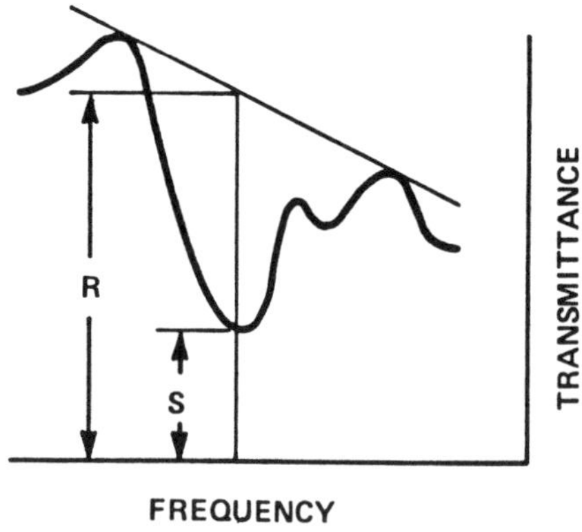

Figure 19: Base line density method of quantitative analysis.

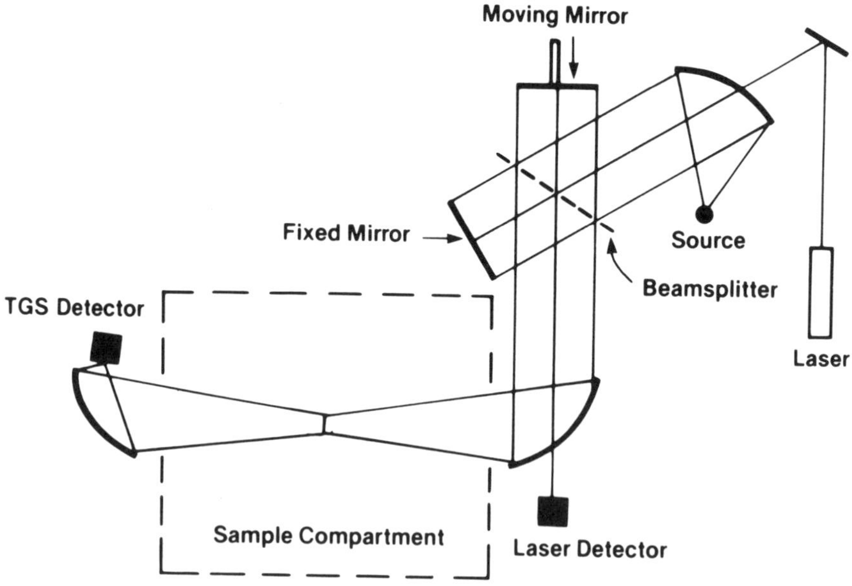

Figure 20: Schematic of Qualimatic S-100–Digilab Division of Bio-Rad Laboratories.

This instrument is able to carry out the following analysis:

(1) Carbon/oxygen in silicon wafers.

(2) Boron/phosphorus in silicon dioxide films.

(3) Epitaxial silicon thickness measurement, n/n^+ and p/p^+ wafers.

The latter capability is valuable for production process control, since a thickness measurement can be completed every 5 seconds. As noted earlier, there is no simple way to evaluate the thickness of such films.

7.3.5 Surface Spectroscopy

In contrast to the previous section where we considered infrared radiation that passed through a thin film and was partially absorbed, we now consider emission from a thin film when we bombard it with radiation. Therefore, we now consider a variety of excitation sources and an equally large number of emitted or backscattered particles or photons. For excitation sources, we have:

(1) Electrons

(2) Ions

(3) Photons (x-rays, UV, visible, IR)

The backscattered or emitted species can be:

(1) Electrons

(2) Ions

(3) Photons (x-rays, UV, visible, IR)

(4) Neutrals

For all of the techniques to be discussed, the process has to be carried out under high vacuum conditions. For this reason, they are not useful for routine process control.

7.3.5.1 ESCA: In this technique (electron spectroscopy for chemical analysis), a beam of low energy x-rays (e.g., the K_{alpha} line of aluminum at an energy of 1.487 keV) is used to bombard a sample. The x-ray photon energy removes an outer shell electron from an atom when it is absorbed. This electron is emitted with a kinetic energy characteristic of the difference between the x-ray and the binding energy of the electron. The energy of the emitted electron defines the type of atom, and the number of electrons at this energy is related to the number density of atoms present.

A typical complete ESCA system including pumps, computer, x-ray power supply plus the sample chamber is shown in Figure 21 (a). Details of the system are illustrated in Figure 21 (b).

In general, photon (x-ray) bombardment of thin films is less damaging to the surface. For this reason, it is possible to obtain information on the top several atomic layers by this process. In fact, not only can we determine the elements present, but we can also obtain information on their bonding. For example, in Figure 22 we show the ESCA spectrum for two polymer samples. One is contaminated with Na and Cl. By examining the Cl peak, it can be identified as being present as a chloride. As another illustration, in examining an aluminum surface, one finds that electrons emitted from aluminum metal atoms have different energies than those emitted from aluminum atoms bonded

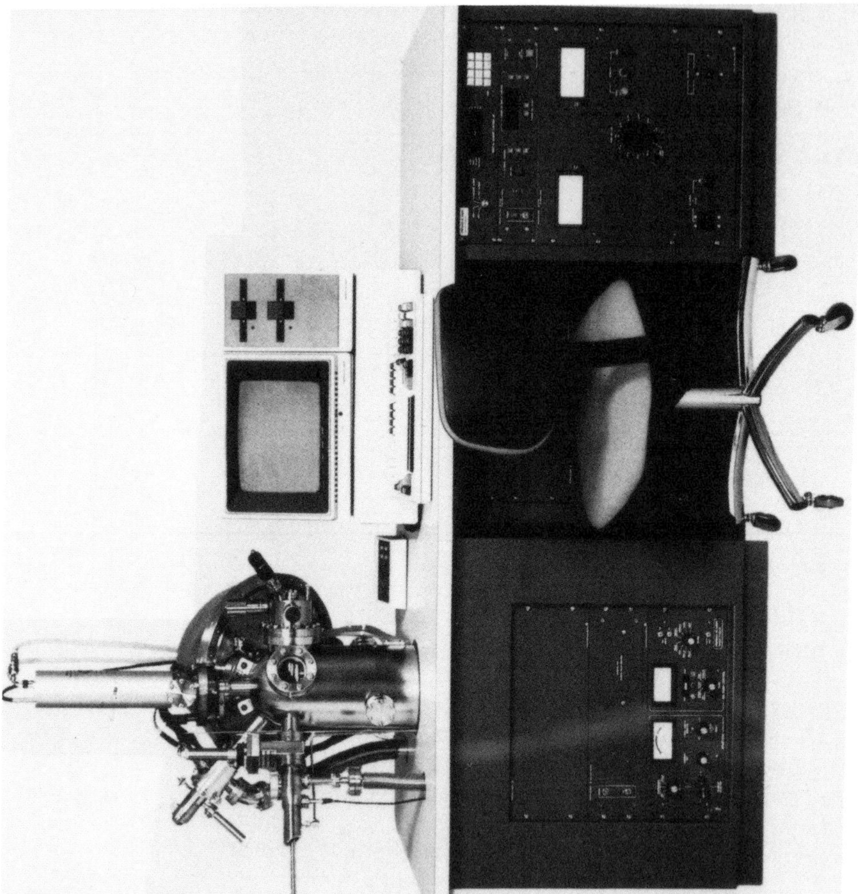

Figure 21: Phi 5000 ESCA system—Perkin-Elmer.

(a)

(b)

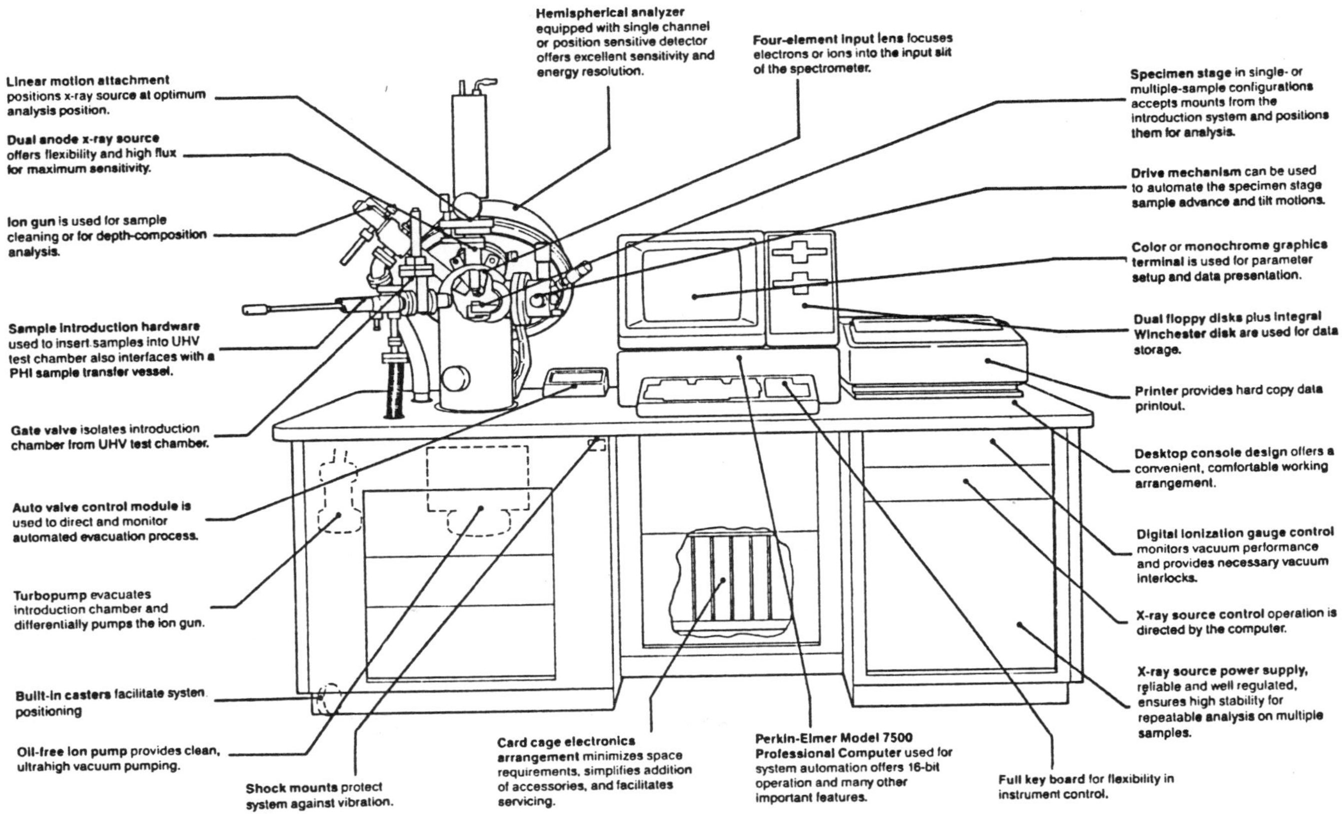

Figure 21: (continued)

to oxygen (aluminum oxide). A typical ESCA spectrum for an aluminum surface is shown in Figure 23. The Al(o) peaks correspond to aluminum metal and the Al(III) to aluminum oxide.

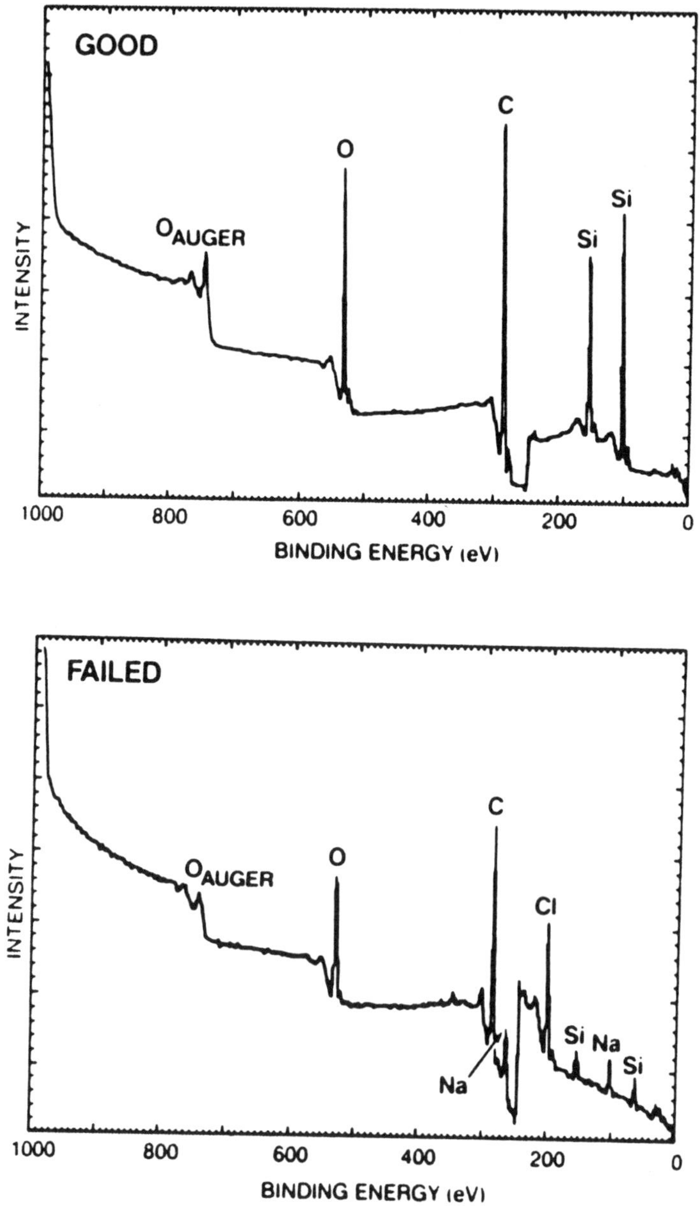

Figure 22: ESCA spectra for clean and contaminated polymer films.

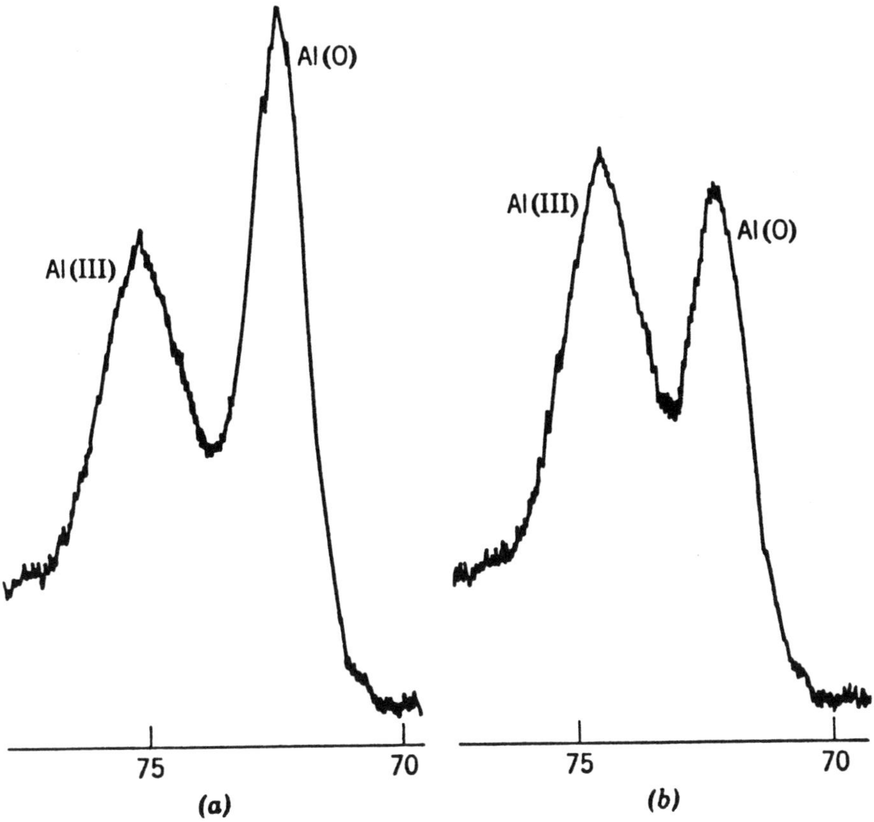

Figure 23: ESCA spectrum of aluminum surface showing metallic aluminum and aluminum oxides: (a) freshly cleaned sample, (b) after 5 days air exposure.[13]

There are several limitations to the use of ESCA analysis. For one, due to the difficulty of focusing x-rays, ESCA has generally been done over several mm diameter samples. Recently, several manufacturers have introduced new systems that can examine areas as small as 150 to 200 microns in diameter, but there is no way one can look at micron-size features. In addition, ESCA cannot detect hydrogen or helium. Finally, it is not a very sensitive process with detection limits of 0.1 to 1.0 atomic percent.

Compared to the other techniques that will be reviewed later, this technique excels in its ability to examine the top several monolayers of a thin film and to offer chemical bonding information, as contrasted to only elemental data.

7.3.5.2 Auger: Another popular technique for thin film analysis is called Auger electron spectroscopy (AES). In this technique, an energetic electron beam (up to 10 keV) is used to probe a surface. The energetic electron can ionize an atom by dislodging an inner core electron. When an outer core electron falls to the inner core to replace the ejected electron, the atom can give up its excess energy by emitting an x-ray. Alternately, it can eject a second "Auger" electron. The energy of this second "Auger" electron is typical of the atom it came from, allowing the elemental composition to be determined. Although the electron beam will penetrate many angstroms into the sample, only Auger electrons produced in the top 10 to 30 Å of the film will contribute to the signal. Therefore, it is truly a surface analysis tool.

If a sputter ion gun is included in the Auger system, one can do a depth profile through a thin film by sputtering to a given depth and then doing the Auger analysis. Also, since the probe is a beam of electrons that can be focussed sharply (~350 Å diameter), the system can be used to do microanalysis of micron-size structures. In this way, an Auger system can be used to provide three-dimensional information on thin film structures by operating it much like a SEM.

The data obtained from an Auger spectrometer is generally presented as the derivative with respect to energy of the number of electrons emitted, N(E), versus energy, as shown in Figure 24. Using the derivative of N(E) rather than N(E) itself helps distinguish the Auger peaks from the significant background. The removal of carbon and oxygen by cleaning the molybdenum silicide surface can be clearly seen.

Depth profiles are usually presented as atomic concentrations versus sputter time, assuming we know the rate at which the sample sputters. A typical depth profile is shown in Figure 25. It is interesting to see that at the surface there is carbon, silicon dioxide and some molybdenum. As soon as the surface layer is sputtered off (300 Å), the oxygen and carbon impurities drop to constant and small values. For this CVD film, the molybdenum silicide came out to be very silicon rich. We can also see that the stoichiometry of the silicide changed with position (depth) in the film.

As with ESCA, Auger analysis can detect all elements heavier than helium. The quantitative presentation shown in Figure 25 can only be done when suitable standards are available. Since the Auger spectra depend secondarily on how the atom is bound in the thin film, it is not always possible to secure accurate standards so that absolute determination of elemental concentrations can be inaccurate. Also, the Auger technique is not very sensitive, and in this regard, is similar to ESCA. Concentrations of an element less than 0.5 atomic % cannot be predicted reliably.

7.3.5.3 SIMS: When ions are used as the probe beam, and they are energetic enough to sputter ions from the surface being studied (500 eV to 5 keV), the secondary ions can be analyzed in a mass spectrometer to indicate the elemental character of the surface. This technique is then called secondary ion mass spectrometry (SIMS). Although the concept is simple enough, there are many practical difficulties in implementing such a process.

Both oxygen and cesium ion beams are used. The former is more effective with the electropositive elements (i.e., B, Al, Cr, etc.), and the latter with the electronegative elements (i.e., C, O, As, etc.). Since the technique involves sputtering of the surface, depth profiling can be carried out readily.

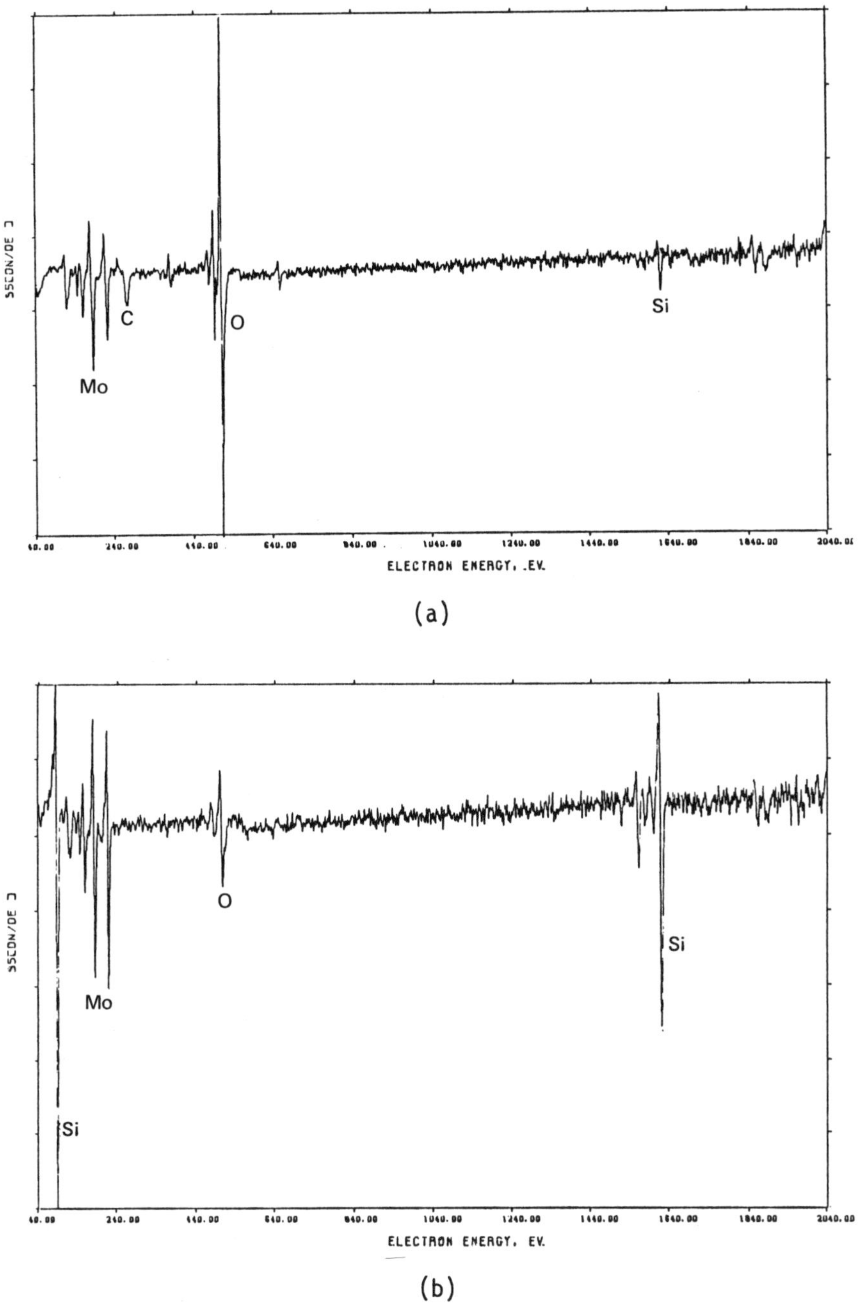

Figure 24: Auger spectrograph of molybdenum silicide film both (a) before and (b) after sputter cleaning.

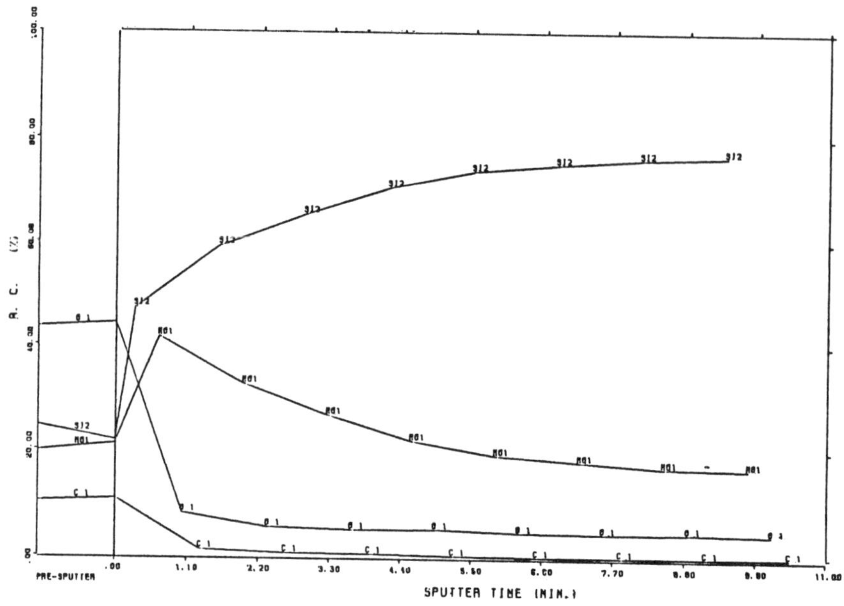

Figure 25: Auger depth profile of molybdenum silicide film.

The single most unique characteristic of the SIMS technique is its sensitivity. It can be as good as one part per billion (ppb). For example, if silicon is sputtered at a rate of 10 Å/sec over an area of 100-μm x 100-μm, then 10^{-11} cm^3/sec of material is removed. Given the density of silicon, this reduces to approximately 5 x 10^{11} atoms/sec. If 1% of these atoms are ionized (by charge transfer with the surface) and 10% of those ionized are collected in the mass spectrometer, then the measured ion intensity will be 5 x 10^8 ions/sec. If we assume we can distinguish 5 ions/sec, then a detection sensitivity of 1 part in 10^8 is achievable. This sensitivity is many orders of magnitude better than other techniques.

Although in principal SIMS can detect hydrogen, this evaluation is not very reliable. The evaluation of hydrogen concentrations will be discussed in the last section. Due to the wide variation in sputtering secondary ion yields, detection limits on SIMS can vary from 10^{-9} to 10^{-3} depending on the species being studied. The technique is most sensitive to those elements with low ionization potential (i.e., Na, K, etc.).

The biggest problem that occurs with the SIMS technique is masking of the species of interest by molecular ions of the same molecular weight. For example, when studying phosphorus doping of silicon, the mass spectrometer will see ^{31}P as well as $^{30}Si^{1}H$, since there can be hydrogen either in the silicon or in the chamber as an impurity. One way to eliminate the $^{30}Si^{1}H$ signal is to recognize that actually there is a difference of 0.008 in molecular weight between these two species. Then, a high mass resolution mass spectrometer can be used to eliminate the spurious signal.

A similar problem occurs with arsenic-doped silicon. Now, the interfering molecular ion is $^{29}Si^{30}Si^{16}O^+$, which has a molecular weight of 75, the same as arsenic. In this case, use is made of the fact that the atomic ion has a broader energy spectrum than the molecular ion, as shown in Figure 26. We can then bias the sample (i.e., –30 to –50V) so that only higher energy ions are detected. Although this reduces the ion intensity of the atomic ions, it has the beneficial effect of eliminating interference from the molecular ions.

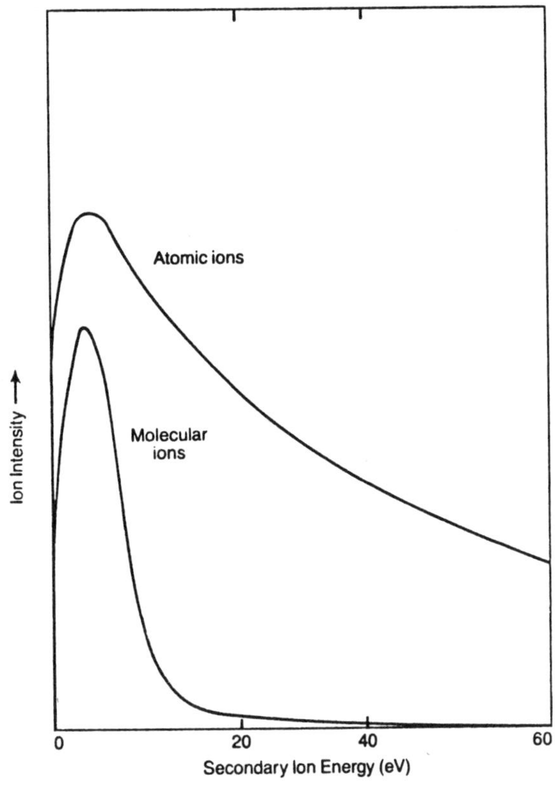

Figure 26: Secondary ion energy distribution.

In the semiconductor industry, SIMS has been particularly useful for the depth profiling of dopants that are present in silicon in very low concentrations. As an example, a SIMS depth profile for boron implanted into silicon is shown in Figure 27. One of the significant features is that we can detect about 10^{15} boron atoms/cm^3 in a silicon matrix of 5×10^{22} atoms/cm^3. This illustrates an ability to detect 20 ppb. Also, the method spans 5 orders of magnitude in boron concentration. No other technique can span such a large range accurately.

Similar depth profiles are routinely done for phosphorus and arsenic-doped silicon films.

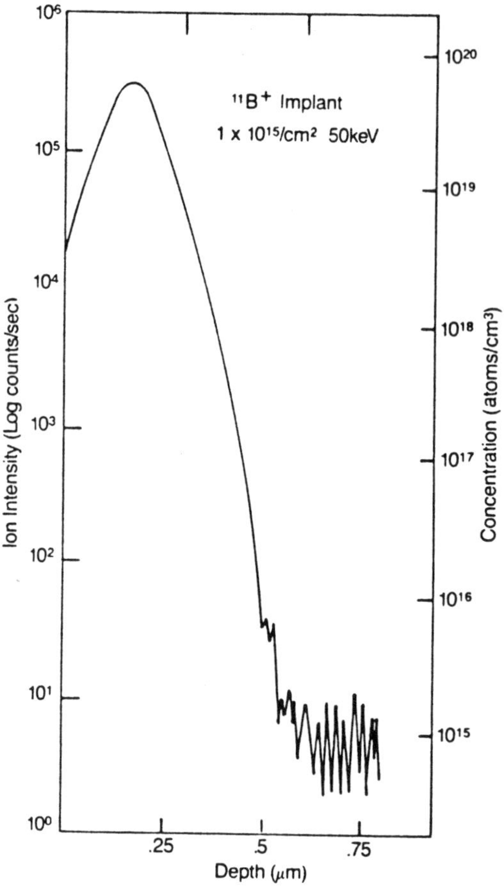

Figure 27: SIMS profile of boron implanted into silicon.

Finally, since it is possible to readily focus the probe ion beam, it has proven feasible to scan with SIMS and display images showing the elemental distributions. One example of such an analysis is shown in Figure 28.

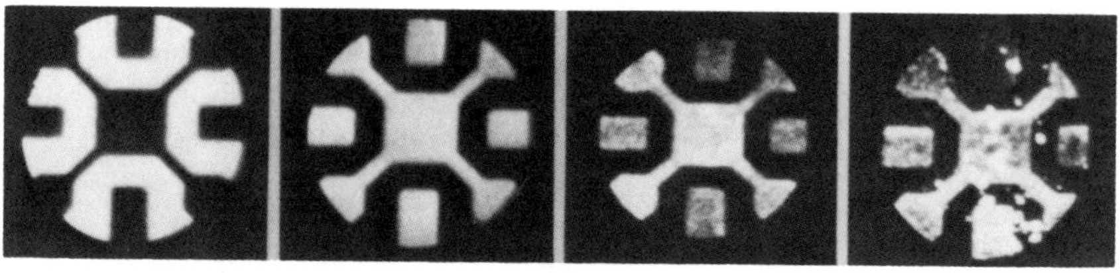

Figure 28: SIMS images showing, left to right, Al, Si, Na in a good device, and Na in a bad device.[14]

7.3.5.4 RBS: There is one surface analysis technique that can give quantitative information on elemental composition and depth profiling without the use of standards. This technique is Rutherford backscattering (RBS). The physical concept behind RBS is quite straightforward. When an energetic ion recoils from an elastic collision with an atom (no angular deflection), its energy depends on the mass of the atom it hit. Such a collision is shown schematically in Figure 29. Typically a doubly-ionized helium ion is used in the probe beam.

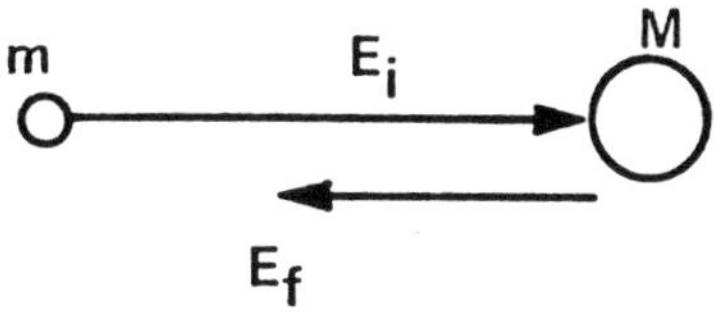

Figure 29: Rutherford backscattering.

The backscattered energy is simply

$$E_f = E_i \frac{M-m}{M+m} \quad , \tag{12}$$

When using the He^{++} ion accelerated to 2 MeV, the RBS procedure is most sensitive to the heavier atoms. A typical detector will have an energy resolution of ~20 keV, regardless of energy detected. Then, when the backscattered ion has an energy close to 2 MeV, the experimental accuracy will be approximately 1%. An example showing several RBS spectra is shown in Figure 30.[15] Here we can clearly see the greater sensitivity of the process for heavier atoms. Also, the probability of a scattering collision occurring is larger for heavier atoms. So the number of counts detected, for the same dose, is substantially higher for platinum than for titanium.

A second feature of RBS is its ability to give concentration versus depth. As the He^{++} ion traverses the sample, it loses energy because it undergoes many grazing collisions before it hits an atom head-on and recoils. Similarly, on its way back through the sample, it again loses energy. This energy loss can be used to evaluate the depth at which the collision occurred. An example showing this depth profiling capability is illustrated in Figure 31. Here, antimony was implanted into aluminum at room temperature. It had a maximum concentration at 750 Å. The second curve shows that it took a 500°C anneal to cause diffusion of the antimony.

Returning to Equation (12), we recognize that the backscattered energies of heavy atoms will be quite close to each other, and it will be difficult to separate them. For example, consider a beam of He^{++} with 2-MeV ions. Then

$M = 50\ m$	$E_f = 0.961 E_i$
$M = 46\ m$	$E_f = 0.957 E_i$
$M = 2\ m$	$E_f = 0.333 E_i$
$M = 4\ m$	$E_f = 0.600 E_i$

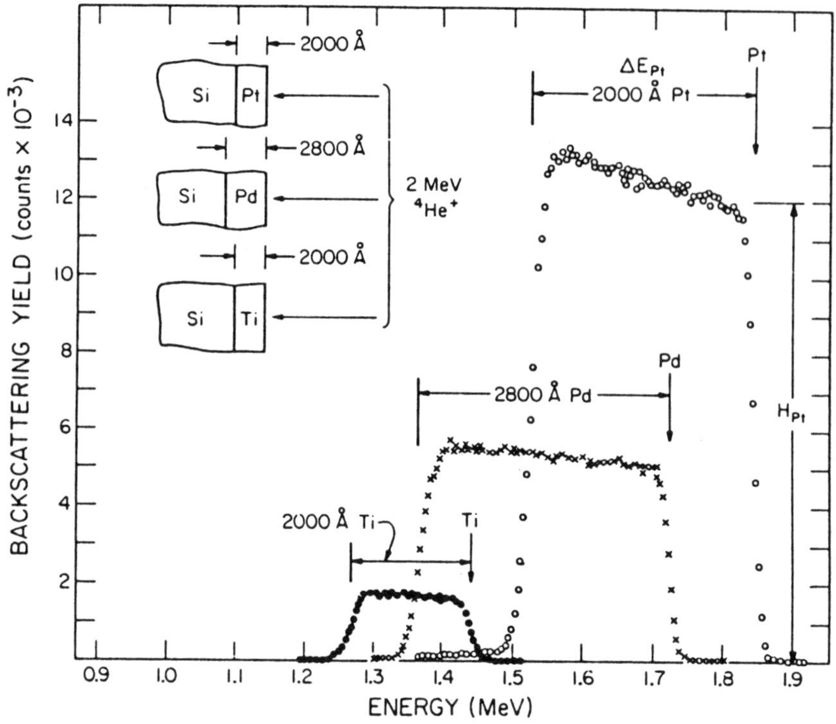

Figure 30: Spectra of 2.0 MeV He^{++} ions with constant exposure to helium beam for each film.[15]

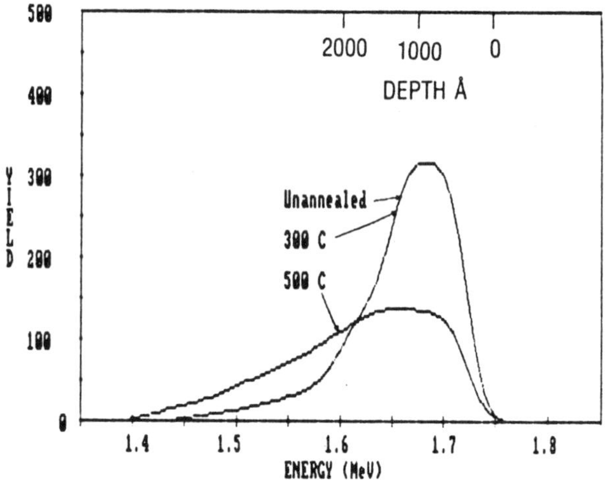

Figure 31: RBS spectra and depth profile for aluminum with implanted antimony.

Clearly, one can distinguish easily between two light atoms, but it will be almost impossible to distinguish between two heavy atoms.

This fact provides guidance as to what systems can best be analyzed by RBS. In general, it will be reasonable to see heavy atoms in a light atom background (i.e., as in Si). Two lighter atoms can generally be distinguished provided there is enough sensitivity to detect them. For example, silicon and oxygen can be seen readily. Finally, detecting a light atom in a heavy atom background is generally possible.

Finally, we must note that RBS gives no information on chemical bonding, so it may be necessary to use another technique to establish the nature of any compound being studied. Also, the beam diameter is typically 1 mm, so one cannot use this procedure to examine small structure in the same way that we can with Auger and SIMS.

Nonetheless, RBS is unique in that it can provide quantitative elemental composition data without resorting to the use of standards.

7.3.6 Hydrogen Concentration Evaluation

In all of our earlier discussions, we have not found any satisfactory technique available to evaluate hydrogen concentration in a thin film. For plasma-enhanced CVD dielectric films, at least, there can be as much as 25 atomic % hydrogen in the film, and its concentration is definitely of interest.

One technique that has been able to measure hydrogen concentration in a thin film and do a depth profile, without reliance on standards, uses a resonant nuclear reaction technique.[16] In this procedure, the nuclear reaction between a hydrogen atom (^1H) and an energetic nitrogen-15 atom (^{15}N) is used. That is

$$^{1}H + {}^{15}N \rightarrow {}^{12}C + {}^{4}He + \gamma\text{-ray} \quad , \tag{13}$$

and the γ-ray comes off at 4.43 MeV. To probe for ^1H, the sample is bombarded by ^{15}N ions and the yield of 4.43-MeV γ-rays is measured. This reaction has a strong resonance when the ^{15}N impacts the proton at 6.385 MeV, and this fact can be used to obtain the depth profile, as shown in Figure 32.

If the ^{15}N beam energy is set exactly at 6.385 MeV, we will observe the hydrogen at the surface. By increasing the ^{15}N beam energy above 6.385 MeV, it will penetrate the film and begin to lose energy. When its energy drops back to 6.385 MeV, another burst of γ-rays will be observed, and their frequency will indicate the hydrogen concentration within the layer. The position within the layer at which the ^{15}N beam energy will drop back to 6.385 MeV can be calculated from available energy loss data.

Using this technique, the hydrogen content of two PECVD silicon nitride films was measured, and the results are shown in Figure 33. The depth resolution obtainable was 50 to 100 Å and is sensitive to better than one part per thousand.

The principal drawback with this procedure is the difficulty in finding the expensive and sophisticated accelerator necessary to achieve the >6 MeV ^{15}N ion energies. Clearly, it is not a production procedure.

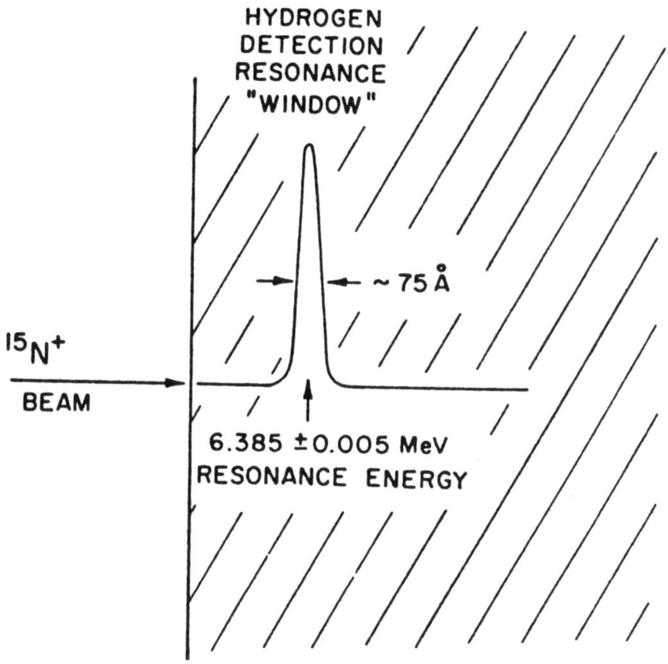

Figure 32: Technique for profiling hydrogen in thin films.[16]

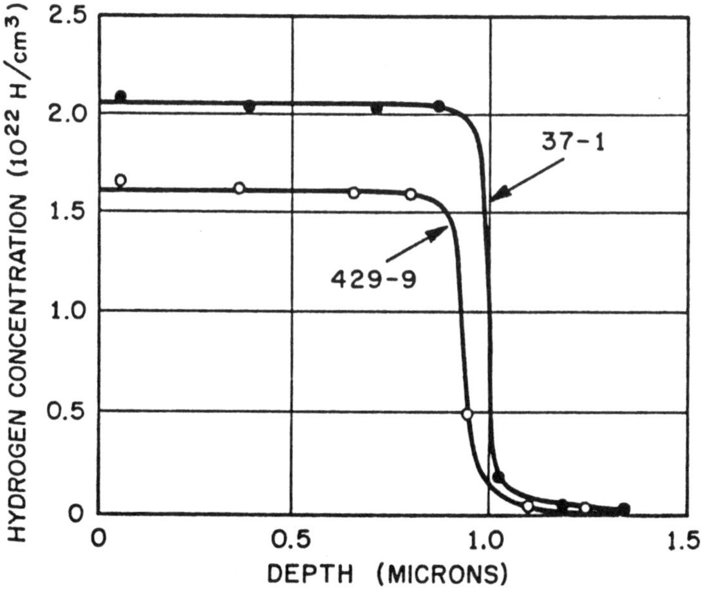

Figure 33: Hydrogen concentration versus depth measured by nuclear resonance.[17]

In order to provide a more convenient method for the evaluation of hydrogen concentration that could be useful for routine process control, the nuclear reaction procedure has been used to provide a standard for an infrared spectrometric technique.[17]

Here the infrared spectra of a thin film on a silicon wafer is obtained first. Then the absorbance at the Si-H and N-H bond regions is measured by estimating the area within each absorption band. As shown in Figure 34, the absorbance seen by the infrared spectrometer correlates well with the hydrogen concentration.

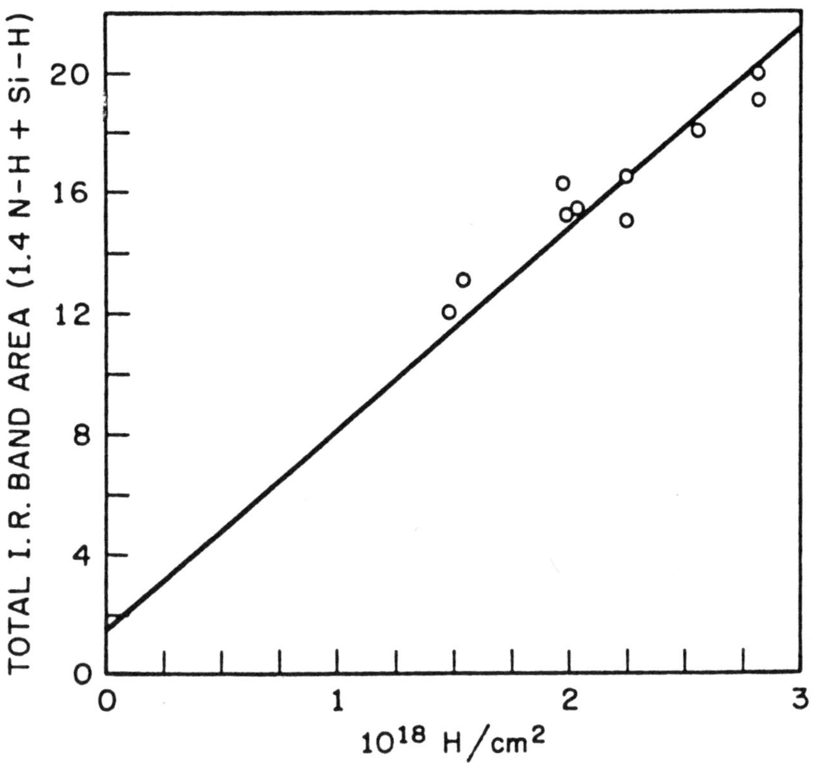

Figure 34: Calibration curve for infrared band area versus hydrogen concentration.[17]

Another technique which can be used to measure hydrogen concentration uses a 2.5-MeV ^{4}He beam.[18] Although this is a more attainable energy range (comparable to accelerators used for RBS) than the nuclear method requires, it is still nowhere near as simple as obtaining an infrared spectra. However, it does offer the capability of depth profiling which the evaluation of infrared spectra does not.

REFERENCES

1. Pliskin, W.A. and Conrad, E.E., Nondestructive determination of thickness and refractive index of transparent films. *IBM J. Res. Dev.* 8:43 (1964).
2. Spanier, R.F., Ellipsometry–A century old new technique. *Industrial Research* (September 1975).
3. Glang, R., Holmwood, R. and Rosenfeld, R., Determination of stress in films on single crystalline silicon substrates. *Rev. Sci. Instr.* 36:7 (1965).
4. Kern, W., Schnable, G.L. and Fisher, A.W., CVD glass films for passivation of silicon devices: Preparation, composition and stress properties. *RCA Review* 37:3 (1976).
5. Smits, F.M., Measurement of sheet resistivities with the four point probe. *Bell Syst. Techn. J.* 37:711 (1958).
6. Zrudsky, D.R., Bush, H.D., and Fassett, J.R., Four point sheet resistivity techniques. *Rev. Sci. Instr.* 37:885 (1966).
7. Jenkins, M.W., A new preferential etch for defects in silicon crystals. *J. Electrochem. Soc.* 124:757 (1977).
8. Linder, R., Bryson, C. and Bakale, D. Surface analysis in semiconductor fabrication. *Microelectronics Manufacturing and Testing* (February 1985).
9. Hilibrand, J. and Gold, R.D., Determination of the impurity distribution in junction diodes from capacitance-voltage measurements. *RCA Review* 21:245 (1960).
10. Gordon, B.J., On-line capacitance-voltage doping profile measurement of low-dose ion implants. *IEEE Trans. on Elec. Dev.* ED-27:2268 (1980).
11. Matsuo, S., and Kiuchi, M., Low temperature deposition apparatus using an electron cyclotron resonance plasma. *Proc. Symp. on Very-Large-Scale Integration Science & Techn.,* Electrochemical Society, N.J. 7:79 (1982).
12. Miller, R.G.J. and Stace, B.C. (eds.), *Laboratory Methods in Infrared Spectroscopy.* Heyden & Son, NY (1972).
13. Adamson, A.W. *Physical Chemistry of Surfaces,* John Wiley & Sons, NY, Fourth Edition (1982).
14. Ward, I.D. and Strathman, M., Analysis methods complement each other in surface studies. *Ind. Rev. Dev.* (September 1983).
15. Nicolet, M-A and Chu, W.K., Backscattering spectrometry. *American Laboratory* (March 1985).
16. Lanford, W.A., Tracetvetter, H.P., Ziegler, J.F. and Killer, J., New precision technique for measuring the concentration versus depth of hydrogen in solids. *Appl. Phys. Lett.* 28:566 (1976).
17. Lanford, W.A. and Rand, M.J., The hydrogen content of plasma-deposited silicon nitride. *J. Appl. Phys.* 49:2473 (1978).
18. Bordin, T.T., Pronko, J.G. and Joshi, A., Quantification of hydrogen in surfaces and thin films using a non-destructive forward scattering technique. *Thin Solid Films* 119:429 (1984).

Index